P9-DIB-589

Energy Efficiency:

Principles and Practices

Energy Efficiency:

Principles and Practices

Penni McLean-Conner

Disclaimer: The recommendations, advice, descriptions, and the methods in this book are presented solely for educational purposes. The author and publisher assume no liability whatsoever for any loss or damage that results from the use of any of the material in this book. Use of the material in this book is solely at the risk of the user.

PennWell Corporation
1421 South Sheridan Road
Tulsa, Oklahoma 74112-6600 USA

800.752.9764
+1.918.831.9421
sales@pennwell.com
www.pennwellbooks.com
www.pennwell.com

Marketing Manager: Julie Simmons
National Account Executive: Barbara McGee

Director: Mary McGee
Managing Editor: Stephen Hill
Production Manager: Sheila Brock
Production Editor: Tony Quinn
Book Designer: Susan E. Ormston
Cover Designer: Kelly Cook

Library of Congress Cataloging-in-Publication Data
McLean-Conner, Penni.
Energy efficiency : principles and practices / Penni McLean-Conner.
p. cm.
Includes bibliographical references and index.
ISBN 978-1-59370-178-9
1. Energy consumption--United States. 2. Energy conservation--United States. I. Title.
HD9502.U52M392 2009
333.79'160973--dc22

2008043266

Printed in the United States of America

1 2 3 4 5 13 12 11 10 09

To my dad, John, who challenged me to write something every day, and to my mom, Cathy, who made sure what I wrote was spelled and punctuated perfectly.

To my husband, Nick, who provides love and support in all of my endeavors.

And to McLean and Tyler, who give me hope, joy, and purpose.

Contents

Part Two: Deliver EE to Consumers

Part Three: Optimize EE Performance

Foreword

Energy and energy use today are on the minds of consumers, policy makers and service providers. The drivers for this attention are concerns for environment, for national security, and for rapidly rising energy costs. One of the best investments that can be made to address energy and energy use is energy efficiency (EE).

This book has been needed for a long time. It gives a complete description of what EE is and how it works, from the broad overview needed to get corporate and government decision makers to adopt and approve programs to the detailed guidance on how to make programs work.

Ms. Conner is a utility executive who understands and communicates the big picture about why EE is good for the environment and for the economy, appreciating as well how a successful EE portfolio will benefit the utility that implements it seriously. As she points out, success in running a EE program requires commitment—knowing not only how to do it but why it is important. This book is built around developing this deep understanding and fostering excitement about what utilities can do.

Many utilities are being asked by regulators or customers to establish robust EE programs, and many more are likely to face this in the future. In addition, there will be growth in EE service providers, who will have a keen interest in understanding the fundamentals in order to develop EE technologies or services. There will also be many people in regulatory positions that will want to understand EE alternatives. Anyone contemplating how to respond should read this book.

— David B. Goldstein
Energy Program Director, Natural Resources Defense Council
Author of *Saving Energy, Growing Jobs*

Preface

Energy seems to be on the mind of everyone these days—from my neighbor down the street who just completed a home energy audit, to the newly formed town energy committee, to regulators, legislators, energy efficiency (EE) program administrators, and energy service companies. The interests vary depending on the perspective. Local neighbors and citizens want a way to combat rising energy costs and help the environment. Regulators and legislators want to form policy that creates the support structure for a sustainable energy future. Program administrators are interested in how to create or rapidly expand programs to meet the increased demand for EE solutions. Energy service companies are interested in developing new technologies and services to support implementation of EE solutions. This book is designed to be a primer for the many stakeholders interested in EE solutions.

Program administrators, whether they are expanding existing EE programs or building brand-new ones, must create a EE culture that is grounded on a solid business case. A successful culture understands how programs and portfolios of programs move through the EE life cycle to ultimately achieve market transformation. Policy creates the framework by which EE is funded and operated. A good policy framework sets the stage for sustainable EE investments by resolving the inherent conflict between the utilities' incentive to increase sales and society's goal to increase end-use efficiency.

Delivering EE involves understanding the market and designing effective programs that are valued by the market. Hence, excellent programs are based on a strong understanding of the targeted audiences, including information on their current energy profiles and end-use applications, as well as how they value investments in energy solutions. Successful program design uses communication channels and delivery channels that reach the targeted audience. The types of programs delivered to the residential and commercial audiences include not only energy efficiency but also demand response and distributed generation.

Finally, successful program administrators are never satisfied with program performance. Rather, they are always looking for ways to improve or optimize programs. Active involvement in organizations that have missions to advance EE is an excellent way to learn new ideas. Evaluation of programs and portfolios is used by program administrators to validate the program achievements and also identify opportunities to improve program performance. With the landscape of EE changing so rapidly, savvy program administrators are also keeping an eye on the future and ensuring that they are well positioned from a people, process, and technology perspective to succeed in the long term

Acknowledgments

There is a thrill in authoring a manuscript—quickly followed by dread of the daunting reality of researching, organizing, and composing over 200 pages of material that will be of value to readers. Over the past 10 months, many colleagues in the industry, along with my family, have shared my journey and helped me to gather research, frame content, and edit chapters. There are many to whom I owe a debt of thanks and gratitude for helping me make this book a reality.

As with my last manuscript, I asked a few close colleagues to serve as my sounding board. These folks helped to shape the content, organization, and flow of chapters. I want to thank this very special sounding board for their valuable review and feedback. They have my deepest gratitude, as I know how many hours they put into providing input. Lisa Shea, Frank Gundal, Charlie Olsson, and Suzanne Farrington, all energy efficiency experts responsible for NSTAR's energy efficiency programs, were at the core of my sounding board. Dave Tomlinson (Progress Energy), who helped me on my first book, was there for me in this effort also. I am fortunate that well-known industry efficiency experts served on my sounding board, including Marc Hoffman (Consortium for Energy Efficiency) and Sue Coakley (Northeast Energy Efficiency Partnership).

I want to thank the many subject-matter experts in the areas of distributed generation, demand response, energy efficiency, evaluation, and market assessment who came to my aid to support the development of content within the chapters. The chapter on distributed generation came together thanks to input from Jack Griffin (DMJM Harris), Fran Cummings (Massachusetts Technology Collaborative), and Arthur Webb (Duke Energy, retired). Bill Mayer (Comverge), who helped me in my first book, came to my aid in the chapter on demand response, as did David Olivier (NSTAR). Rebecca Foster (Consortium for Energy Efficiency), Roseanne Brusco (NSTAR), and Margaret Norton (NSTAR), who helped with the development of content on market outreach and assessment. Monica Nevius (Consortium for Energy Efficiency) provided insight for the chapter on evaluation, and Nelson Medeiros, Jan Gudell, Barry McDonough, and Patrick McDonnell (all of NSTAR) provided valuable insight that enriched the content of the chapters on energy efficiency.

Many colleagues were quick to respond to my request for interviews and quotes. To all of these folks, I want to express my sincere gratitude for the input and ideas. Thanks to Mike Sullivan (PEPCO), Ted Schultz (Duke Energy), Harvey Michaels (M.I.T. Energy Initiative), Trevor Lauer

(DTE Energy), Bruce Anderson (Wilson Turbo Power), Steve Cowell (Conservation Services Group), Sam Krasnow (Environment Northeast), Bud Vos (Comverge), Bob Laurita (ISO New England), Doug Bastion (North Atlantic Energy Advisors), Meg Matt, (Association of Energy Service Professionals), and Carl Blumstein (California Institute for Energy and Environment).

In addition, I want to thank Steve Hill and the team at PennWell for providing me with this opportunity and for finalizing the manuscript into print.

Thanks are also owed to several folks behind the scenes who helped to make the manuscript complete. The tremendous artwork in the manuscript is thanks to Jim Connelly (NSTAR). Mike Durand (NSTAR), who is always ready to proofread, came to my aid several times by reviewing the content and flow of various sections. Luann Jencyowski (NSTAR) served as my right arm; she played many roles and was pivotal in the creation of all the tables, the organization of material, the development of artwork, and the review of copy.

My family reminded me at all times to keep this effort in perspective. Both boys, Tyler and McLean, made sure that I did not get too serious and that I stopped to play, and my husband, Nick, provided ongoing encouragement and support throughout.

When a book comes out in print, the moment of truth rests with the readers. I feel confident, thanks to the many folks who helped me along the way, that readers will find this book valuable in helping them to advance EE.

Part One:

Create an Energy Efficiency Culture

Chapter

Build the Business Case for Energy Efficiency

Many of us remember the gas lines of the 1970s, when fuel prices hit all-time highs. Society responded by advocating for change. As a result, building codes were enhanced, appliance standards were adopted, many Americans moved to more fuel efficient cars and homes, and early energy efficiency (EE) programs began.

In the early 1980s, EE incentives were first implemented to accelerate advancement of energy efficiency programs. In 1988, the National Association of Regulatory Utility Commissioners (NARUC) passed a resolution urging regulators to "make the least cost plan the utility's most profitable resource plan."[1]

When the wave of competitive energy supply swept the country, the belief prevailed that the competitive market would offer energy efficiency services, thus avoiding the need for regulatory intervention in the form of rates to fund and utility incentives to offer EE services. Indeed, as Trevor Lauer, vice president of marketing for DTE Energy, remarked, "Once deregulation swept through the country, energy efficiency wasn't so cool; deregulation was the sexy piece. Everyone felt it was better to save 10 to 15 percent on a kilowatt-hour, versus saving kilowatt-hours."[2] Today, 10 years since the beginning of competitive energy supply, competitive offerings have not materialized. As a result, policy makers are renewing discussions on how to expand energy efficiency.

Today the concern about energy and the environment has reached unprecedented levels. Unlike the fuel crisis of the 1970s, which created short-term passion for energy conservation, there is a real concern—and a sense of urgency—to protect our global environment not only for our generation but also for our children's generation. Interest in and awareness of potential climate change impacts is at an all-time high, powered by messages in the popular media and by political debate.

The forecast is that within 25 years, our nation's population will have grown by 25% and electricity usage by 40%.[3] Most of this usage, over 70%, will be consumed in our homes, business, schools, governments, and industry.[4]

To meet the needs of the future, investments need to be made now. A natural first place to address growing energy needs is EE. EE is essentially about getting more value from our energy and teaching society to use less energy without giving up the comfort and economic advantages energy provides.

One only needs to look to California to see the impact of sustained EE efforts. California has pursued energy efficiency for over 30 years. As a result per capita electricity consumption in California has remained flat, as compared with a continuing rise in consumption for the United States in total (fig. 1–1).[5]

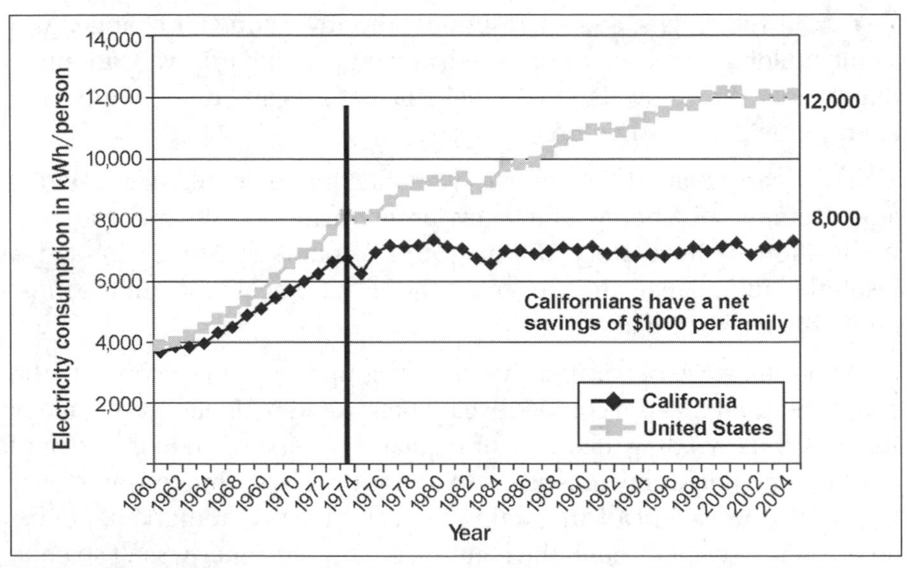

Fig. 1–1. Per capita electricity consumption—California versus the entire United States (Source: U.S. Department of Energy, Energy Information Administration [http://www.eia.doe.gov/emeu/states/sep_use/total/csv/use_csv])

EE is a cost-effective approach to meet our increasing energy needs and can be balanced with supply options. EE today can be provided at approximately one-third of the cost of supply—3.2¢ versus 8.9¢ per kilowatt-hour (kWh) for generation in an analysis by the Massachusetts Division of Energy Resources.[6] Creating and sustaining investment in EE requires commitment from policy makers, regulators, utilities, and other stakeholder groups and requires building a business case for change. Elements of that business case include defining EE, understanding the benefits, and identifying the barriers. Additionally, utilities will play a significant role in EE regardless of whether they are also serving as program implementers.

What Is EE?

EE is an investment, just as a transmission line or a power plant is. More precisely, it is an investment in reducing kWh to meet rising energy demands. Energy efficiency has a focus on reducing energy waste. In the words of Edward Vine of Lawrence Berkeley National Laboratory, "Historically, policy makers and utility regulators have considered EE [energy efficiency] as the least cost strategy to help meet resource adequacy and transmission expansion needs, especially in geographically congested areas."[8] EE reduces greenhouse gas emissions by decreasing consumption and peak demand, thereby delaying or avoiding capacity upgrades.

EE is delivered to consumers through program administrators. These most commonly are utilities, but other organizations, including state agencies, municipal organizations, or third parties, also administer programs.

Program administrators are accountable for designing cost-effective EE portfolios that meet the overall state energy goals or strategies, along with internal objectives. These goals and objectives will specify energy savings targets, but also often include societal-related goals, such as reductions in greenhouse gas emissions or the requirement to design programs that equitably serve all customer classes.

EE portfolios are made up of an integrated set of cost-effective programs serving all customer classes, ranging from residential and low income to commercial and industrial. Designing and managing a portfolio that maximizes energy benefits and meets all objectives is both an art and a science. Successful program administrators work with their teams to create a robust portfolio that is vetted with stakeholder groups and eventually is filed for approval by appropriate regulatory bodies. Once a plan is approved, administrators focus on managing the delivery of the programs. Administrators work with an internal team, along with external partners and energy service providers, to implement residential, commercial, and industrial programs.

Energy efficiency tops the list of programs delivered by administrators. Today, though, program administrators are expanding their programs to include other demand side management (DSM) efforts, such as demand response, dynamic pricing, and distributed generation. Each of these has its own unique technology, design and delivery processes and implications for utilities and customers. Some measures, such as energy efficiency, have an impact on both kWh and kilowatt (kW) demand, while others, such as demand response, primarily have an impact on kW demand.

Energy efficiency programs comprise residential, low-income, commercial, and industrial programs that focus on end-use technologies. Program administrators strive to move technologies forward with the goal of achieving market transformation. Market transformation occurs when consumers buy and use the most efficient technologies as a part of their normal routine. To advance energy efficient technologies, program administrators identify the impact of market barriers of awareness, availability, accessibility, and affordability and design a program to mitigate those barriers.

Demand response is defined as "changes in electricity consumption by customers in response to signals in the form of electricity prices, incentives, or alerts during periods when the electricity system is vulnerable to extremely high prices or compromises to reliability."[9] Demand response programs add temporary electricity capacity on very short notice for spikes in load or short-term deficiencies in supply by providing incentives to customers to reduce load on receiving notification. There are two forms of demand response: incentive-based and time-based rates.

Incentive-based demand response can be offered to residential and commercial customers. Commercial customers respond to a notice by either curtailing use by turning off lights and air handlers or shifting production schedules among other measures or turning on self-generation. Residential customers can also participate through controls on air conditioners, water heaters, or pool pumps. Demand response is managed and owned by utilities or demand response providers, who act as an interface between the independent system operator and utility and the end-use customers to deliver demand response capacity.

Time-based rates involve establishing energy prices that vary based on the time the energy is provided. This type of approach is intended to affect consumer behavior by sending appropriate price signals. When prices are high, the consumer benefits by reducing load. Time-based pricing generally moves load from a peak to a nonpeak time, versus eliminating load, like energy efficiency. There are several forms of time-based pricing, including time-of-use pricing, critical peak pricing, and real-time pricing.

Distributed generation is the use of small-scale power generation technologies located close to the load being served. Interest in and installation of renewable distributed generation (e.g., solar and wind), along with higher-efficiency distributed generation (e.g., combined heat and power), are becoming more prevalent, especially as states adopt policy that encourages the investment in renewable distributed generation.

Benefits of EE

EE and other DSM efforts benefit consumers, businesses, and the organizations that provide the services from an economic perspective. Further benefits include environmental, economic development, and energy security.

Lowers energy bills

Perhaps the most commonly cited benefit is the savings consumers gain from investing in EE. Consumers gain more than just an economic benefit. They also gain control and understanding of their energy usage. When a customer calls a utility with a high-bill complaint, often the underlying issue is the consumer's perceived lack of control over the cost. Education and investment in EE will provide consumers with the ability to gain control of their energy usage.

The U.S. Environmental Protection Agency (EPA) notes that over 30% of energy usage is wasted. Indeed, all consumers—residential, commercial, and industrial—have opportunities to save energy. The Alliance to Save Energy has noted that if over the next 15 years Americans bought only Energy Star products, energy bills would shrink by more than $100 billion.[10] According to Energy Star, a program managed by the EPA, if each U.S. home replaced just one of its incandescent lightbulbs with a compact fluorescent lightbulb (CFL), the electricity saved each year could light three million homes, and over $600 million in annual energy costs would be saved.

Business and industry also gain from investments in EE and DSM solutions. Investments in energy efficiency for this sector typically have short paybacks, particularly if combined with program incentives, and provide a business with a competitive edge.

Represents cost-effective investment

Perhaps not as evident to consumers is that the investment in EE will also have an impact on the future cost of energy. As Sam Krasnow, a policy advocate for Environment Northeast, observed, "We have a choice, we can spend eight cents on supply or invest three cents on efficiency."[11] The choices to date have been to invest in supply (see fig. 1–2). Spending on generation supply in New England alone is 40 times the investment in efficiency.

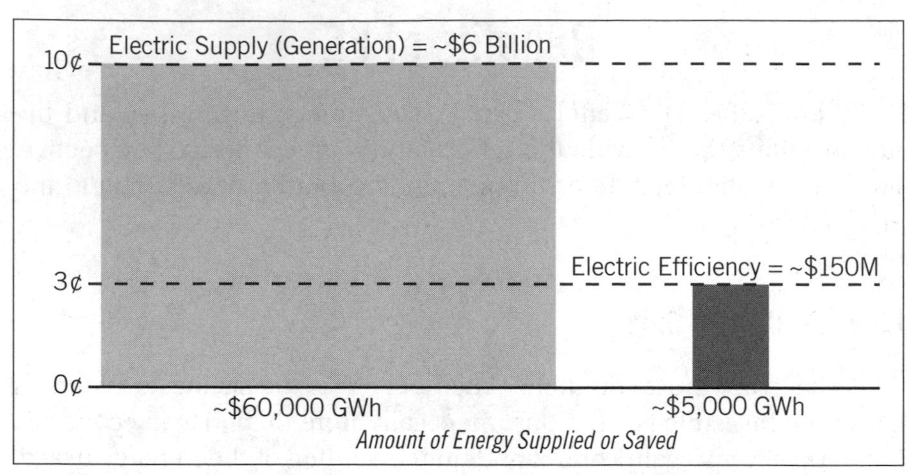

Fig. 1–2. Comparison of electric spending on supply versus efficiency (Source: Environment Northeast)

Increasing the investment in EE and DSM delays the need for investment in additional generation. This translates into even more energy savings for consumers.

Achieves fast and significant energy savings

EE is the fastest way to address growing energy demands. Efficiency programs can be scaled and implemented in a short period of time, often in one to three years.[12] The programs can also be targeted to areas where growth is outpacing the ability to bring in new capacity from a generation, transmission, and distribution perspective. Well-designed energy efficiency programs are delivering annual savings on the order of 1% of electricity and natural gas sales.[13]

Delivers environmental benefits

Fossil fuels produce emissions that have an impact on our climate and our air quality. Electricity production, of which over 50% is powered by coal, is a leading source of emissions (fig. 1–3).

EE programs that reduce demand for energy thereby also reduce greenhouse gas emissions. In addition to emission reductions and energy savings, EE also brings benefits of lower water use and less environmental damage from fossil fuel extraction. These additional benefits are commonly referred to in the industry as *nonenergy benefits*.

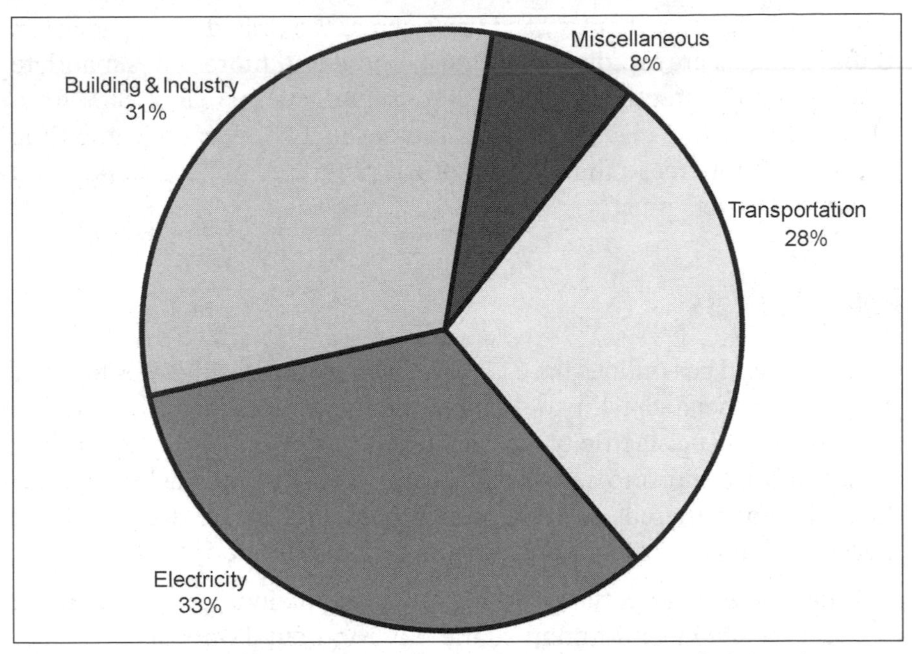

Fig. 1–3. Sources of emissions (Source: Massachusetts Technology Collaborative [http://www.masstech.org/cleanenergy/important/envemissions.htm])

Fosters economic development

Simply put, EE is good for the economy. Companies that invest in energy efficiency upgrades find themselves in a more competitive position. There is job creation in states that pursue energy efficiency. Trevor Lauer of DTE Energy has attested that EE "is huge for economic development. It builds jobs."[14] In Massachusetts, for example, an annual growth rate of 20% is expected in industries related to clean energy. The largest sector of this industry is jobs associated with energy efficiency and demand response, representing 44% of the sector.[15] Additionally, participants in energy efficiency programs and services often redirect their bill savings toward other activities that benefit the local economic with higher employment impact than if the money had been spent to purchase energy.[16]

Barriers to EE

Investment in EE is funded inconsistently among the states in the United States. A few states represent a majority of the funding, with the top 15 states representing 80% of the nationwide spending on energy efficiency programs. There is good news across the United States, however, in that budgets are rapidly increasing as more and more states mandate funding for EE programs. Gas efficiency budgets increased 68% from 2006 to 2007 while electric budgets increased 14% over the same time interval.[17] To increase the funding of EE programs policy, market and customer barriers must be removed.

Policy barriers

Electric and gas utilities have long focused on increasing sales to cover the costs associated with the expensive investment in infrastructure needed to deliver electricity and gas to homes. Not so long ago, it was common for utilities to sell appliances as a way of increasing load on the system and thereby to more quickly gain a return on the significant investment in generation, transmission, and distribution.

Today there are important state, regional, and national goals to increase end-use efficiency and minimize the environmental impacts of energy production and consumption. This presents a problem for utilities as it is not in a utility's best interest from a sales perspective to help customers to reduce energy usage, thereby reducing revenue. Healthy utilities need that revenue to ensure access to attractive capital, to continue the necessary investments in maintaining and expanding the system.

Policy changes are needed to create and sustain investment in EE. It is through policy that rate-making frameworks can be developed that create a positive environment for utility investment in EE. Policy also identifies and mandates funding for EE efforts and ensures that building energy codes and appliance efficiency standards are adopted and implemented.

A rate-making framework is required that addresses the inherent conflict between implementing EE programs and the need to increase utilities sales. Common frameworks include decoupling, modified rate structures, fixed customer charges, and loss-based revenue. Each of these frameworks attempts to make the utility whole for losses associated with the implementation of EE programs. Complementing these frameworks are incentives for the utility to aggressively pursue EE. Incentives are necessary to allow EE to compete with other utility investments such as transmission and distribution, both of which earn a healthy rate of return for shareholders.

Stable and multiyear funding is needed to support advancement in EE. Some states have successfully provided stable funding through a system benefits charge, which is a charge, on the customer's bill, that funds energy efficiency. Other funding mechanisms in play include rate-based recovery mechanisms and resource procurement funding, each of which has benefits and drawbacks.

Building energy codes and appliance efficiency standards complement ratepayer-funded demand-side programs by locking in the market gains of programs that successfully build the availability and use of new technologies and best practices. This regulation frees up funding for voluntary programs to build the market for the next tier of cost-effective demand-side resources.

Market barriers

To achieve success, EE efforts must bridge market barriers. There are four broad categories of market barriers: awareness, affordability, accessibility, and availability. These barriers affect customers' adoption of EE efforts and must be overcome by program administrators to successfully delivery cost-effective energy management solutions.

Awareness. Awareness is a barrier when customers lack information on available options. Customers may not know how to differentiate between various energy management options or even be able to recognize the substantial energy savings benefits that an investment in energy efficient technologies can provide.

Customers for whom English is not the first language also face an awareness barrier. These consumers are often unaware of energy programs that may be readily available. Because organizations providing the energy management services are always challenged to keep the costs of the programs low, there is a reluctance to spend the extra dollars needed to reach out to non–English-speaking or other hard-to-reach customers.

Availability. Availability is a barrier when manufacturers either do not make or do not effectively market and produce significant quantities of the energy efficient products they do make. Availability may be affected by the lack of the manufacturer's or industry partner's foresight to aggressively develop a market for their products. Although program administrator incentive mechanisms are intended to accelerate emerging markets, there are times when incentives can become disconnected from an end-user or value proposition perspective.

An example of this is the split incentive. This situation occurs when the beneficiary of the energy improvements is not the decision maker, as when a builder wants to construct a home or building cost-effectively, to quickly turn a profit on the investment. Builders are often hesitant to spend the extra dollars required in order to upgrade insulation and heating, ventilation, and air-conditioning (HVAC) systems because they are concerned about the price impact. They do not benefit from the investment. As a result, only 10% of new homes are built to the Energy Star level.

Accessibility. Accessibility refers to customers' access to the product. It involves distribution retailers' stocking product, or displaying products in sufficient quantities and in visible areas so that consumers can find them. Conversely, while there may be plenty of products available, the supporting infrastructure—such as contractors or installers with sufficient technical skills, experience, and certifications needed to successfully promote, sell, and properly install energy management solutions—may be lacking.

Affordability. Typically, energy efficient solutions are more expensive than the products they are meant to replace. More efficient products are often less developed and have a smaller market. A common example is CFLs. Not only do these lightbulbs have a new, snow cone shape, but they are also more expensive than incandescent lightbulbs. Hence, today, over 15 years since these bulbs were introduced, a typical house has CFLs in less than 20% of applicable sockets.

High upfront costs to make the needed upgrades are also a barrier for consumers. Program administrators offset this barrier through a comprehensive portfolio of incentive mechanisms to motivate customers to maximize energy saving opportunities by making proactive investments in energy efficiency upgrades, as opposed to waiting for end of life-cycle equipment failures. While energy service companies do a great job of offering incentives, explaining the benefits, and detailing the payback, consumers are often reluctant to make the initial investment. An example of this is the home energy improvement program offered in Massachusetts, where less than 20% of residential customers qualifying for weatherization improvements actually complete the improvements; this is despite a program that pays for 50% or up to $1,500 of the cost of the measure and coordinates the actual installation.

Utility Role

Regardless whether a utility is currently providing some EE services, societal interest in addressing global warming has implications to the utility. States like Massachusetts that have long-standing and very successful energy efficiency programs are evaluating policy and funding changes that would support increasing the effort dramatically. Other states where no programs existed are now mandating programs. In fact, a broad coalition of utilities, public service commissioners, and businesses, along with the U.S. Department of Energy and the EPA, developed a National Action Plan for Energy Efficiency. This plan, published in 2006, "presents policy recommendations for creating a sustainable, aggressive national commitment to energy efficiency through gas and electric utilities utility regulators and partner organizations."[18]

For utilities, now is a time of great change and opportunity. In talking to utility executives who are actively engaged in expanding or establishing EE efforts, common elements emerged that are needed for success. These are described in the sections that follow.

Top-level support and vision

Capitalizing on this change and redefining the utility in this new environment takes vision and top-level support. Jim Rogers of Duke Energy embodies this vision by speaking passionately on the need for change and has personally dedicated himself to defining how this change should happen. As co-chair of the National Action Plan for Energy Efficiency, Rogers testified before the U.S. Senate that "a genuine paradigm shift is necessary if we are to realize the full potential of this resource. That shift must occur in the way regulators treat the business of energy efficiency, in the way utilities develop and deliver programs, and in the way in which we appeal to consumers to manage their energy use."[19]

Harvey Michaels, a leading expert in energy efficiency and the former chief executive officer of Nexus Energy, asserted that such top-level support and vision "requires an investment of time by senior management of the utility to develop trust in political and regulatory stakeholders that they are sincerely committed to demand side management."[20] Michaels emphasized that leaders like Rogers spend time on this topic as evidenced by their public actions—including testifying, speaking, and actively working on industry efforts such as the National Action Plan for Energy Efficiency.[21]

Create a business case

EE must make business sense for the utility. Ted Schultz, vice president of energy efficiency at Duke Energy, remarked that "you have to figure out how you create a business model that works for customers, the utility, and the environment. This involves creating a business line around energy efficiency so that it compares favorably to building a generation plant from a financial perspective."[22]

This business case is important not only for external audiences, like regulators and stakeholders, to establish the appropriate rate-making framework, but also for internal audiences, to align employees around the effort. In fact, a change-management effort that features the business case and communicates it to employees at all levels is often used by companies to create alignment within the organization. Changing long-established processes and mind-sets that have supported a model of increasing sales to favor a business model of reducing load is a significant undertaking.

Organize to deliver

As utilities prepare to take on a significant role in reducing energy usage, they must be organized to deliver this change. For companies starting EE efforts, selecting leadership, defining processes around the EE life cycle, and establishing partners is important. Some companies with established EE programs are facing similar challenges as they organize to deliver significantly more EE services as a result of significant increases in EE funding.

As program administrators organize, they need to consider the integration of energy efficiency into the core functions of the business. This applies in particular to utility-managed programs, where integration into customer service, engineering, and operations is a must, in order to successfully deliver programs. To achieve the most success, EE must be integrated into the long-term resource-planning model for generation, transmission, and distribution planning.

Communicate

Communicating with customers, stakeholders, and employees is fundamental as a utility pursues EE. The business case forms the foundation for communications that build support for consumers, stakeholders, and employees to invest time and resources in advancing EE.

Stakeholder communications are also a key element. Stakeholders can play a critical role in the design of rules and regulations associated with EE. Ensuring that stakeholders understand the issues and positions of the program administrator is important. This is where becoming involved in regional and national organizations that advocate policy is beneficial. These organizations, such as the Alliance to Save Energy and the Northeast Energy Efficiency Partnership, provide value by reaching out to policymakers and influential stakeholders in preparation for and during regulatory or legislative proceedings or forums at the state or federal level to address generic policy issues. These organizations also will research and prepare communication materials that aid in advancing discussions on appropriate regulatory structures that support expanded EE.

In the case of employee communications, the establishment of a business case for change is the foundation. A change-management communications plan can be developed to effectively communicate the business case to all levels of employees, as appropriate.

Customer communications are fundamental to the success of EE programs. The most successful program administrators integrate energy efficiency communications and information into all customer touch points, including the call center, World Wide Web, and customer bill.

Summary

A EE culture is founded by building a business case for investment in EE. As issues related to global warming receive more press and the country's focus on conservation increases, the opportunity for EE becomes even greater. Utilities, along with other providers, will play a significant role in delivering EE programs. Building a business case for investment in EE is important for all parties, especially utilities, regulators, and consumers.

A successful business case starts with a definition of EE. This definition can center on energy efficiency but may also include DSM efforts like demand response and distributed generation. The business case should help stakeholders to understand why EE is necessary; therefore, it needs to document the benefits that EE brings both to consumers and to program administrators. There are many barriers that will surface when launching EE programs. A business case documents the barriers and the development of mitigation plans addressing those barriers.

Program administrators across the country have created or are creating a EE culture. The EE culture is based on a business case that resonates with all stakeholders and serves as the foundation on which EE programs can be designed, delivered, and maximized.

References

1 Statement of Position of the NARUC Energy Conservation Committee on Least-Cost Planning Profitability (July 26, 1988), as cited in Steven M. Nadel, Michael W. Reid, and David R. Wolcott, eds. 1992. *Regulatory Incentives for Demand-Side Management.* Washington, DC: American Council for an Energy-Efficient Economy, p. 25.

2 Lauer, Trevor. Interviewed by Penni McLean-Conner on February 29, 2008.

3 Munns, Diane. 2007. Models of efficiency. *Electric Perspectives*. July/August: 20.

4 EPA. July 2006. National Action Plan for Energy Efficiency, p. ES-1.

5 Rosenfeld, Arthur. 2007. Empowering customer to combat climate change. Presented at the Nexus Energy Software Client Conference. Scottsdale, AZ.

6 Bowles, Ian; Massachusetts Backs Efficiency, Energy Biz; September/October 2007. pg. 108.

7 U.S. Department of Energy, Energy Information Administration. Electric Utility Demand-Side Management 1999. http://www.eia.doe.gov/cneaf/electricity/dsm99/dsm_sum99.html

8 Vine, Edward. 2007. The integration of energy efficiency, renewable energy, demand response and climate change: Challenges and opportunities for evaluators and planners. Presented at the Energy Program Evaluation Conference, Chicago.

9 California Public Utilities Commission. 2007. Order Instituting Rulemaking, Rulemaking 07-01-041 (January 25, 2007). San Francisco: California Public Utilities Commission.

10 Center for ReSource Conservation. Energy Efficiency 101. http://www.conservationcenter.org/e_energyefficiency101.htm

11 Krasnow, Sam. Interviewed by Penni McLean-Conner on February 12, 2008.

12 EPA. July 2006. National Action Plan for Energy Efficiency, p. ES-4.

13 EPA. July 2006. National Action Plan for Energy Efficiency, p. ES-4..

14 Lauer, 2008.

15 Global Insight Inc. 2007. *Massachusetts Clean Energy Industry Census*. Report prepared for the Massachusetts Technology Collaborative Renewable Energy Trust, p. 1.

16 Kushler et al., 2005; New York State Energy Research and Development Authority, 2004; National Action Plan for Energy Efficiency, 2006, p. ES-4.Kushler, M., Ph.D., York, D., Ph.D., and Witte, P., M.A. (2005, January) *Examining the Potential for Energy Efficiency to Help Address the Natural Gas Crisis in the Midwest*. Washington, DC: American Council for an Energy-Efficient Economy [ACEEE]. New York State Energy Research and Development Authority [NYSERDA] (2004, May). *New York Energy $martSM Program Evaluation an dStatus Report to the System Benefits Charge Advisory Group, Final Report*. Albany

17 Consortium for Energy Efficiency. 2008. CEE 2007 Report: Energy Efficiency Programs, pp. 8–9.

18 EPA , July 2006. National Action Plan for Energy Efficiency, p. ES-1.NAPEE, 2006, p. ES-1.

19 Edison Electric Institute. 2007. Duke Energy CEO Rogers calls for "paradigm shift" to realize full potential of energy efficiency. News release, February 12. http://www.eei.org/newsroom/press_releases/070212.htm

20 Michaels, Harvey. Interviewed by Penni McLean-Conner on January 24, 2008.

21 Ibid.

22 Schultz, Ted. Interviewed by Penni McLean-Conner on February 10, 2008.

Chapter 2

Understand the EE Life Cycle

The EE life cycle at the highest level involves design, implementation, and evaluation. Successful program administrators not only understand the EE life of specific programs but also integrate programs into a portfolio to achieve overall energy savings objectives.

A group of EE programs is referred to as a *portfolio*. The objective of a portfolio is to accelerate the adoption of efficient energy-use practices and technologies. In a successful EE portfolio, the specific programs are managed over their life cycle to effectively meet the desired strategic objectives. These objectives can be both societal and organizational and are often multidirectional. The management of a portfolio is also characterized as *iterative, flexible,* and *variable,* as it must adapt to changes in program goals, customer adoption rates, technology availability, and implementation availability.

This chapter provides an overview of market stages, an understanding of which is fundamental to successfully manage a program through the EE life cycle. Additionally, the steps in the life cycle and the integration of programs into a portfolio are discussed, to provide a comprehensive overview of EE life-cycle management.

Market Stages

Efficient energy-use technologies and practices move through market stages. These market stages are similar to the stages—introduction, growth, and maturity—through which any new product progresses. Ultimately, the success of a technology or practice occurs when the market transforms or matures. Reaching maturity means that a program or technology has overcome all of the market barriers of awareness, affordability, accessibility, and availability. Program managers, when designing programs, must evaluate the maturity of the specific measure or technology (table 2–1).

Table 2–1. Market stages

Stage	Characteristics	Barriers
Introduction	Limited number of customers Few market channels	Awareness Affordability Accessibility Availabilty
Growth	Increasing number of customers Increasing number of channels	Affordability
Maturity	Supported by code Ubiquitous	None

When a technology is first introduced, the market is immature and is characterized by early adopters. Early adopters are inclined to test new technologies and even help with the technology evolution. Early adopters in the EE world are often referred to as *efficiency pioneers*. To participate in the introduction of a program, efficiency pioneers must bridge all the barriers of awareness, availability, accessibility, and affordability. Efficiency pioneers must be convinced to accept a higher cost and take a risk on performance.

Adding to the challenges of managing this market stage are the limited availability of the technology, the few market outlets, and the very few product features. Research and development may be needed during this market stage, as a technology emerges. The data on performance of the technology are limited, as are the data on customer acceptance. The barriers to customer adoption include customer awareness, availability, accessibility, and affordability. In an immature market, all of these barriers arise, making the role of the program administrator particularly challenging.

Consider the challenges faced by program managers at the advent of the CFL during the early 1990s, when it was introduced into the marketplace. Trying to convince consumers to purchase a radically new, snow cone–shaped lightbulb at $10.00 a piece to replace a common $0.59 lightbulb based on 100 years of proven technology was an almost impossible feat. Market acceptance of this emerging and unproven energy-saving technology was further compounded by limited availability, limited uses and sizes, lack of consumer demand, and higher first costs at a time when energy costs were relatively inexpensive.

Program administrators worked individually and collectively—in partnership with Energy Star, using its nationally recognized branding power—to develop strategic initiatives and incentive mechanisms at the manufacturer, distributor, and wholesaler levels. The goal of these initiatives was to increase product inventory availability and introduce competitive pricing at retail outlets. These efforts were also supported by major national and local consumer education and public awareness campaigns promoting the benefits of CFLs and often were aligned with program administrator rebate incentives. As a result, the CFL market has achieved a steady increase in market share with steady decreases in wholesale pricing. Product variety and consumer acceptance continues to evolve and has begun to reach the mature level of self-sustainability. This scenario demonstrates the important role program administrators play in the market transformation of new technologies. It is likely that with light-emitting diode (LED) lighting on the horizon program administrators will play a similar role in the eventual transformation of this market as well.

The next stage in the EE program life cycle is a growing or maturing market. This stage is characterized by an increasing number of accepting customers and many market channels. In a maturing market, the program administrator still must address the affordability barrier, but often in this stage, awareness, availability, and accessibility are not as large a barrier, if they are present at all. Program administrators may continue to use incentives in a maturing market to encourage consumers to pay the additional cost associated with more efficient products and services.

The final stage is a mature market, or market transformation, at which point the technology is commonplace, used by consumers without the need for any additional incentives. A mature market is supported by codes and standards. The technology simply becomes the best option—if not the only option, owing to standards. A mature market has overcome the final barrier of affordability. The technology becomes one that customers choose freely. There are no barriers to awareness, availability, or accessibility at this stage.

To support and determine the level of market maturity, national studies are constantly being conducted, often in conjunction with the EPA and the U.S. Department of Energy to determine actual market shares of various emerging or existing technologies. Program administrators use these studies to evaluate and refine their approaches to advancement of the technology. This is an important element in the design of program, which is the first step in the EE life cycle.

EE Life Cycle

The life cycle of a specific program encompasses design, implementation, and evaluation, all supporting the objectives defined in the overall EE program portfolio. Design of the program will involve an assessment of the market, technology, channels, and cost and savings impacts associated with the program. Implementation takes the program design and puts it into practice; this is often accomplished through technology installation and consumer education. Evaluation involves assessing the costs, savings, and consumer impacts.

Design

Program design is both an art and a science. The design involves understanding the market through an assessment of customer and corporate needs and objectives, and then iteratively evaluating potential demand-side options by using a screening process. As options are developed, the design will progress to include the type of program, communication and marketing channels, and delivery mechanism. Additionally, there will be a determination as to whether the technology is to be offered as a stand-alone product or bundled with other technologies (fig. 2–1).

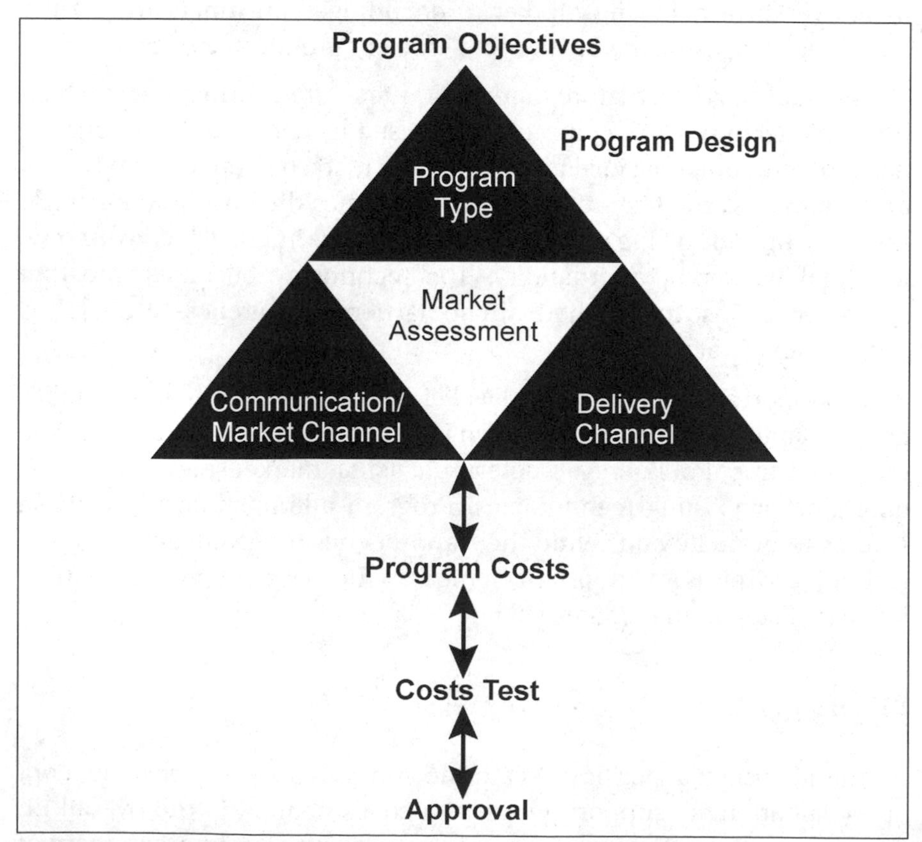

Fig. 2–1. Design process

The first step in the design process is to define the program objectives. Time spent carefully defining the objectives will help in the more fluid part of the process, which is developing the program concept. The program concept involves defining the market segments, assessing the market, determining the delivery strategy, and identifying the costs.

Market segmentation involves determining which customer sectors will be targeted by the program. Market research is often used in this stage. This research may be accomplished by using focus groups or surveys and reviewing the experience of other program administrators who have implemented similar programs. This research will aid in identifying the target market and estimating participation.

The market assessment, which is discussed in more detail in chapter 4, involves understanding load objectives, customer sectors, and end uses involved to meet specific load objectives. Additionally, an assessment of the existing market barriers of availability, affordability, awareness, and accessibility is completed at this juncture. Marketing and delivery strategy is also defined for the program design. In this step, the delivery mechanism is developed, along with tactics to drive demand, such as promotion, incentives, and trade allies.

With the target market identified and assessed, and the delivery mechanism designed, an estimate of the costs can be completed. This estimated cost will be screened for cost-effectiveness, in which likely program scenarios are compared on the basis of different cost options, rebate and incentive levels, and differing levels of utility commitment. Program administrators may use one or multiple tests to screen program designs. These tests, the use of which varies and may be dictated by state regulators, include the total resource cost test (TRCT), the utility cost test, the participant cost test, and the rate impact measure test.

The TRCT includes all quantifiable costs and benefits, regardless of who accrues them. The utility cost test includes only those quantifiable costs and benefits that are accrued by the utility system. This test excludes participants' costs. The participant cost test includes costs and benefits accrued by the program participant. The rate impact measure is the net change in electricity utility revenue requirements. This test attempts to measure rate impact on all utility customers, especially those who do not directly participant in the EE program.

Pilot programs are used by program administrators to test new program measures, customer adoption, and savings impact. Pilot programs are often a great tool for evaluation of new technology, allowing the program administrator to find out, on a small scale, the pros and cons of a specific measure. This information can be incorporated into a full program rollout. The cost-effectiveness tests are more lenient with pilot programs. Pilot programs generally do not run more than 24 months.

The program must go through an approval process. Many utilities use a gate acceptance process that requires all new ideas or programs to pass through a series of approvals prior to receiving final approval. This process ensures that all facets of the project have been considered. This includes program objectives, market segments, and market assessment, market and delivery strategy, and program costs. Clearly, defining the program concept is fluid and both an art and a science. Once the program concept is completed, programs are then combined to review from a portfolio perspective. This is described later in this chapter (see the "Portfolio Management" section).

Implement

With the program designed and approved, it is ready for implementation. At this point, the program manager will determine how to manage the program. This will often involve third-party vendors who deliver engineering, audit, and installation services to residential and business consumers. In an effort to leverage marketing channels, program managers may work with other program administrators, retailers, and Energy Star to market and promote various efficiency measures.

Program managers will use a competitive bid process to hire energy service companies to implement energy efficiency measures. These firms may provide intake, distribution in-field services, and, in some cases, integrated marketing support. There are many highly qualified vendors who specialize in home or business audits and associated installation. Engineering support may also be secured to aid implementation. This is common in the commercial sector owing in part to the complexities of estimating energy savings and deferring financial paybacks as a means to encourage and secure investments in new technologies by property owners and building developers.

Program managers have identified key elements that ensure quality program implementation. These elements are focused on providing a quality customer experience and having timely feedback mechanisms on the implementation process to assess and adjust appropriately.

Ensuring quality fulfillment of the request. Ensuring that customer requests can be fulfilled in a quality and timely manner is important for high customer satisfaction with the program. Program managers should take time to ensure that marketing material accurately provides contact and fulfillment information, that customer calls will be handled in a timely and quality fashion, and that delivery is completed in a timely and quality fashion. Before marketing materials head out the door, program managers should complete a final review and test the instructions provided to the customer. For example, sending out a marketing piece with an incorrect telephone number is frustrating to customers and adds unnecessary expense for the program. Confirming that product availability for consumers is sufficient to meet expected demand requires communication with the distribution company. Ensuring that personnel on-site are installing as planned and in a professional manner is also important.

Ways to ensure quality fulfillment include surveying customers, mystery-shopping, and monitoring performance metrics. Program managers may want to complete transactional surveys, typically within one week of the transaction, to gain the customer's perspective on the service. Mystery shopping involves a staff person impersonating a customer to gain feedback on the fulfillment process. Metrics that may be monitored include cycle time for delivery, call answer rates, and order backlog. It is very important to have feedback mechanisms in place to monitor the fulfillment process to achieve high customer satisfaction with the process.

A program administrator told me about his experience with an executive mystery shopping. The utility executive, on his way home from work, heard a radio commercial for an energy efficiency program and decided to mystery-shop the experience. He immediately called the toll-free number given on the radio advertisement, only to be placed on hold for several minutes and then disconnected. He attempted the call again, with the same experience. In debriefing the situation the next day, the learning point was that fulfillment occurred only during business hours, which ended at 5:00 p.m.; the problem, thus, was that the radio commercials continued to be broadcast into the evening hours. This situation was quickly corrected by extending call-center hours to ensure good service. A habit of mystery-shopping a new fulfillment campaign is a good one to quickly assess the quality of an efficiency campaign.

Ensuring that marketing channels are reaching the desired audience. Program managers always strive to minimize the costs associated with marketing and outreach. As such, it is helpful to identify channels that penetrate the specific audience you are targeting. For example, if print media is being considered, which outlets reach the target audience? If the campaign is localized, like an air conditioner turn-in event, then community newspapers may be a better and lower-cost option than a regional newspaper. With direct mail, can you surgically identify who should receive the mailings?

One innovative program manager designed a direct mail campaign that targeted the customers on either side of a home that had participated in an efficiency program. The mailer highlighted that their neighbors had recently participated in the same program and had taken advantage of the generous rebate offerings listed. This type of mailing, known as a *surgical mailer*, reduced overall program and marketing costs owing to its inherent ability to generate higher response rates as compared with a blanketed-type mailing lacking any type of demographic or prescreening information.

Partnering with other organizations. Partnering with other organizations, such as other program administrators or community-based groups, is an excellent way to cost-effectively implement a program. This can reduce the costs of marketing outreach and improve fulfillment.

For instance, combination of budgets allows for more robust energy efficiency messaging campaigns. The campaigns are run on a much greater scale, as measured by increased media buys and increased messaging frequency, than if an administrator had tried to do it individually.

An example of this is the Energy Bucks program offered to consumers in Massachusetts. Energy Bucks is an integrated campaign combining grassroots outreach, community-based activities, and advertising to build awareness of the variety of energy efficiency and discount services available to income-eligible families. Energy Bucks is sponsored by the local utilities and municipal aggregators in Massachusetts that provide administration for energy efficiency programs, in partnership with local community action program (CAP) agencies and the Massachusetts Department of Housing and Community Development. This program provides participants with one-stop shopping through a common Web site and toll-free number. The CAP agencies provide the intake services, and third-party vendors do the actual audit and installation.

Partnering with Energy Star. Energy Star is one of the strongest partnerships an energy efficiency administrator can have. Like the Good Housekeeping seal of approval of yesteryear, the Energy Star logo has evolved into a nationally recognized symbol that signifies energy efficiency. It has also become the benchmark on which consumers base their trust and purchase decisions when buying new household products and appliances. In fact, data show that 68% of households recognize the Energy Star label.[1] Energy Star was initiated in 1992 as a joint program of the EPA and the U.S. Department of Energy. It is a voluntary labeling program designed to identify and promote energy efficient products to reduce greenhouse gas emissions. The Energy Star program provides the best vehicle for program administrators to maximize consistent program messaging, to influence the standards and testing of consumer product labeling, and to convert its product ratings and recommendations into regional and localized energy efficiency initiatives for consumers.

Complaint resolution process. Whether internal or external (through vendors), there need to be processes to deal with consumer complaints. It is extremely important that customer concerns and complaints are heard and resolved in a timely and effective manner. It is equally important when a customer complaint arises on the vendor side for there to be established two-way communication channels between the program administrator's staff and the vendor to resolve the complaint.

The process itself can be simple but, again, must be timely and effective. The basic process is to act immediately on the complaint. Determine the most effective and mutually agreeable way to resolve it, and document the actions taken. Last—and probably the most important and overlooked step—is following up once a complaint has been resolved. A simple call back to the customer can make a world of difference by bringing closure to the complaint and will often produce a longer-term positive impact on customers' attitudes and opinions toward the company.

Timely fulfillment. Once consumers decide to participate, it is critical to fulfill their requests in a timely manner. The fulfillment may be as simple as mailing out material or as complex as scheduling an installer. Program managers must be in constant contact with their distribution partners to ensure adequate supply and prompt response. Measuring fulfillment cycle time is a good tool, along with assessing the overall quality of the services and products being provided.

One utility discovered a problem with timely fulfillment through customer satisfaction surveys. Rebate processing had the lowest satisfaction of all programs. In talking to and surveying many of these customers, it was found that the time to receive the rebate, which was averaging three months, was longer than customers expected and, in many cases, was unacceptable from a customer satisfaction perspective. To solve this issue, a Six Sigma team was organized with the utility and the distributor to review the entire process. In addition to shortening the rebate time by 75%, they simplified the customer rebate form through a redesign, eliminated unneeded information, and developed new procedural guidelines with the vendor. These enhancements led to a 47% decrease in rebate form rejection rates and a six-percentage-point increase in customer satisfaction within the first three months.

Evaluate

This step measures how well the program met its goals from a process and impact perspective. A process evaluation identifies areas for program delivery improvement. This is often done using surveys and interviews with third-party partners, staff, and customers. An impact analysis measures quantifiable success based on reductions in kW, kWh, or therms. These savings are compared to the original program objectives and are used as a first step in determining savings attributed to the program. Adjustments are then made to the initial or tracked savings estimate. These adjustments are commonly referred to as *realization rates*, or *impact factor*, and are expressed as the ratio of the evaluated savings divided by the tracked savings.

Depending on state regulatory policy, further impact factors are often measured, including an adjustment for free-ridership and spillover. Free-ridership is a measure of how many participants would have purchased the technology without the program. Free-ridership reduces overall savings. Spillover is when a customer implements further measures because of the program but without being offered any additional incentives. An estimate of savings attributable to spillover is then added to the initial savings estimate. The end result is the net savings associated with the implementation of the program.

The evaluation will also make a determination on market saturation. Market saturation is defined as when a market matures to a level of self-sustainability whereby market shares achieve continuous growth through purchases being made based on true customer demand and perceived value, rather than an incentive-based purchasing decision. For example, in a recent evaluation of incentives for front-loading washers, it was determined that 73% of the time, consumers would have purchased the more efficient front-loading washer without the incentive rebate.[2]

The evaluation may also define other benefits, sometimes referred to as *non-energy* benefits. Examples are resource savings, including savings of key resources such as water, and non-resource benefits, including benefits of reduced operation and maintenance costs associated with the new measure. Also, the evaluation will determine benefits, such as the impact on greenhouse gas reductions. A more thorough discussion on the technology, processes, and tools of evaluation is presented in chapter 10.

Successful programs lead to market transformation. Over time, consumers will move from tentatively adopting a new technology to the point at which consumers purchase the more efficient technology nearly exclusively. This transformation is accelerated through EE programs.

Portfolio Management

Portfolios are designed to maximize strategic objectives by using a variety of EE strategies. When EE funding is ratepayer based, regulators will often define strategic objectives. Utilities may also have specific objectives for the EE portfolio that must be incorporated. Program administrators are tasked with developing a portfolio that meets the needs of all stakeholders, both internal and external, and are accountable for delivering cost-effective energy savings and aligning with state energy efficiency goals or strategies.

A well-designed portfolio will meet the defined strategic objectives of regulators and the utility by serving a varied mix of customers through a variety of channels. Utilities should consider the EE portfolio along with other utility-sponsored initiatives. Programs such as the *flat bill,* once considered a strong utility revenue-smoothing program, can be viewed as anti-conservation, thus sending a mixed message to the consumer about the utility's commitment to energy conservation.

Portfolio objectives

EE objectives vary from state to state but often mandate equitable use of funds among customer classes and energy savings targets. Objectives may also encourage adequate support for lost-opportunity efficiency programs, such as new construction, remodeling, or replacement of worn-out equipment. A focus on market transformation, whereby consumers buy and use the most efficient technologies as a part of their normal routine, is the ultimate goal. *Market transformation* is defined as a "reduction in market barriers resulting from a market intervention, as evidenced by a set of market effects that lasts after the intervention has been withdrawn, reduced or changed."[3] Some states may also have specific goals around low-income assistance that must be incorporated.

To meet these objectives—and utility objectives of running a cost-effective generation, transmission, and distribution system—four EE strategies are commonly deployed: energy efficiency, peak load reduction, load shifting, and load building. An organization may focus on one strategy, depending on its overall objectives (table 2–2).[4]

Table 2–2. Utility EE strategies

EE Strategy	Utility Objective	Program Example
Energy efficiency	Reduce overall energy use	Programs promoting high-efficiency end-use equipment
Peak load reduction	Reduce usage at peak times	Air-conditioning system control programs
Load shifting	Permanent load shift to lower use periods	Time-of-use pricing programs
Load building	Increase load during off peak times	Outdoor-lighting programs

An energy efficiency strategy will focus on reducing overall energy use, thereby lowering customers' bills. This strategy often has a focus on end-use technologies and advanced building performance practices. This can involve persuading customers to purchase or use the new technology, influencing market channels to offer the technology, or addressing codes and standards.

A peak load reduction strategy is one where the program focuses specifically on targeted peak load periods. Most common among strategies here is load control for air conditioners or water heaters. Utilities practicing integrated least-cost planning may include a strategy centered on demand reduction, a lower-cost option than building, or adding supply. While such strategies do have some impact on reducing overall usage, the intent is instead to reduce usage during peak time, understanding that this usage may reoccur off peak. For this strategy to be of maximum effectiveness, there needs to be control by the utility or system operator to tap into the load reduction when it is most needed.

Load shifting involves moving load to cheaper times. This strategy will also have an impact on peak load, but it is a permanent move of load to lower-cost times. Often load shifting is complemented and incented through price signals associated with a time-of-use rate or critical peak-pricing rate. An example of a program that focuses on load shifting is the process of using ice storage for cooling. In this situation, refrigeration is used at night to create ice. During the day, the ice is used for cooling.

Load building is yet another strategy and involves encouraging customers to use off-peak energy programs. Utilities looking to increase load factor or increase density on a gas main may target a load-building program. Natural gas efficiency programs often have load building as a strategy and offer incentives for high-efficiency natural gas systems as system replacements for more expensive alternative fuels such as oil.

Portfolio process

Programs are combined to create a portfolio that meets the top-level organizational objectives. Typically, portfolio plans are created annually and submitted to appropriate regulatory bodies for approval. During the course of the year, the portfolio is managed to ensure that the objectives are successfully met. Portfolio management is an iterative, flexible, and variable process that involves several steps (fig. 2–2).

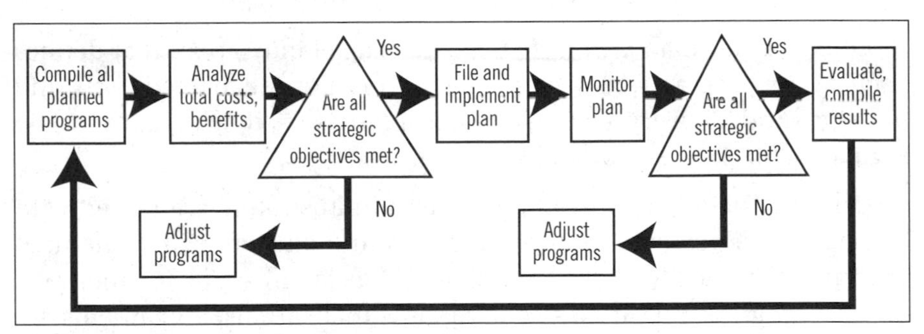

Fig. 2–2. Portfolio process

In portfolio creation, all planned programs must be combined to determine their overall impact. At this stage, programs and their underlying assumptions are refined, along with their specific goals and objectives. In ratepayer-funded programs, a conscious effort is often made to have the funds properly allocated to customer classes. This helps to define the portfolio. Program administrators will also take into account market saturation as a criterion in allocating dollars for programs. Additionally, input and feedback from both internal and external partners is reviewed and incorporated into the programs. From an internal perspective, specific program goals and definitions are analyzed from a portfolio view. Program administrators typically use customized software to review the specific programs and combine them for overall impact.

The second step is to analyze the total costs and benefits. Efficiency funds are normally capped by the amount of funding available. Program administrators always want to maximize their investment to achieve the most savings, while meeting all other strategic objectives. At a minimum, each program must also achieve more savings than costs. Well-designed portfolios will achieve benefits greater than two times the costs. Some portfolios have delivered much higher savings, in the range of four to one or greater. Typically, programs targeting the residential market will have lower benefit cost ratios than commercial and industrial programs. This is due to the high cost of marketing and delivering programs to the mass market.

There is an iterative step of analyzing program costs and reviewing strategic benefits and then adjusting the programs. This is where administrators must make key decisions on which programs to scale back or expand, to achieve total cost and benefit goals, along with all strategic objectives. A variety of both external and internal reviewers may also weigh in on various drafts of the portfolio scenarios. Their input is used by administrators to define the final portfolio plan.

Once the portfolio is finalized, it is packaged into a plan that defines each of the programs, as well as the assumptions, expected costs and benefits, and delivery strategies. This plan is filed with appropriate regulators for their review and final approval.

Once the plan is approved, program administrators are tasked with managing the portfolio. Individual program managers will implement their plans. Typically, there are frequent checkpoints, at least monthly, on plan progress. It is at these checkpoints that various adjustments are made to individual programs or to the portfolio.

At the end of the program year, the program administrators complete a very thorough internal evaluation of the portfolio plan. This evaluation determines specific savings achievements and documents final costs. This is an exacting process, and often incentive dollars are associated with the success of the plan.

A second measurement and evaluation is completed on some programs 12 to 18 months after the end of the program year. The purpose is twofold: to validate the actual savings of that program and for regulatory reporting purposes. Validated savings data are used by program administrators for planning purposes for future programs. Usually, this type of program evaluation is completed by third parties to ensure objectivity. Every program is not evaluated in this manner each year, as this review is considered valid for planning purposes for an extended period of time. Also, this is a more cost-effective approach.

Portfolio administrators must be flexible to adjust to changes in priorities, inaccurate assumptions on demand, and material and vendor availability, along with other changes. When changes occur, effective portfolio management involves reviewing all programs and adjusting goals and targets to ensure the portfolio objectives are achieved. For example, when NSTAR offered a new technology to customers, the demand was 15 times greater than initially expected. This was primarily because of media coverage not anticipated when the initial marketing assumptions were developed. Program decisions had to be made in light of overall portfolio objectives. Options in this case were to ratchet down the demand by closing the program, which, although not an uncommon practice, can result in customer irritation. Another option was to let the program continue and adjust other programs correspondingly. This option has the downside of reducing resources needed for other planned programs. During the course of a program year, a variety of issues will arise requiring adjustment, sometimes minor but sometimes drastic, to achieve overall objectives.

Portfolio management is variable. Program administrators will approach the management task from different perspectives and with different processes and tools. Some organizations use a team approach, while others develop the programs by customer sector and then integrate the portfolio at a higher management level. There is not a standard analytical tool, which will roll up individual program goals into an overall goal for the portfolio. Most of the analytical tools or software packages currently available are homegrown by individual utilities' program administrators and are unique to the specific goals of that organization. The processes by which the portfolio is developed will also vary from program administrator to program administrator and will be refined over time.

Summary

Understanding and effectively applying the EE life cycle is critical in the successful implementation of programs and in the creation of a EE culture. An EE culture is always focused on achieving market transformation or maturity. Program administrators understand the steps of developing programs to move technologies toward transformation. An organization with an EE culture will develop program portfolios that deliver results both in energy savings and in customer satisfaction.

There is an art and a science of portfolio management responsible for delivery of high-value programs to consumers. Successful administrators create a culture in which innovative programs are continually brought forward. These administrators understand the strategic objectives and follow a disciplined life-cycle process to design and deliver programs that meet these objectives. The end result is the acceleration of the adoption of efficient energy-use practices and technologies through value-added programs.

References

1 Environmental Protection Agency, Office of Air and Radiation, Climate Protection Partnerships Division. 2007. *National Awareness of Energy Star® for 2006: Analysis of CEE Household Survey*. Washington, DC: U.S. Environmental Protection Agency, p. ES-1.

2 Nexus Market Research Inc. Estimates of net impact of the 2006 Massachusetts Energy Star appliance program, clothes washers component, p. 18. July 2007.

3 Eto, Joseph, Ralph Prahl, and Jeff Schlegel. 1996. A scoping study on energy efficiency market transformation by California utility DSM programs. Earnest Orlando Lawrence Berkeley National Laboratory, Energy and Environmental Division, p. xii.

4 Johnson, Katherine, and Ed Thomas. 2007. *The Fundamentals of Linking Demand Side Management Strategies with Program Implementation Tactics*. Market Development Group.

Chapter 3

Influence Policy to Support EE Investment

Edison Electric Institute forecasts that the industry will spend more than $73 billion for new plants and infrastructure to keep pace with the country's growing demand for electricity, which is predicted to increase by 40% by 2030.[1] This predicted increase in electricity demand, combined with concern for the climate, has resulted in many states and nations experiencing a paradigm shift in how energy policy makers and utility regulators see the world. The "energy lens" has changed to a "climate change lens," through which energy policies and programs can now be viewed as solutions to the problem of reducing greenhouse gas emissions.[2] The significance of this shift is that it broadens the effort and involves integration of other demand-side strategies, including renewable energy, distributed generation, and demand load control.

Creating and sustaining investment in EE requires commitment from policy makers, regulators, utilities, and other stakeholder groups. Commitment often translates into policy, and policy can have a great impact on advancing energy efficiency technologies and options. Policy that ensures consistent implementation and coordination with building energy codes and appliance efficiency standards can have a significant impact on meeting load growth through demand-side options. For example, the Pacific Northwest region has met 40% of its growth over the past two decades through energy efficiency programs.[3] Sue Coakley, executive director of Northeast Energy Efficiency Partnerships, observed that "Successful energy efficiency programs require a long-term, stable policy commitment to overcome the barriers to achieving the potential for cost-effective efficiency. The most successful public policies . . . integrate efficiency and demand-side resources as a cornerstone of long-term resource plans to meet energy, environmental and economic goals."[4] It is through policy that funding is identified and mandated for EE efforts and that building energy codes and appliance efficiency standards are adopted and implemented.

This chapter provides a high-level overview of the key policy frameworks that advance energy efficiency, including funding mechanisms, rate structures, program administration, and building energy codes and appliance efficiency standards.

Funding Mechanisms

Funding for energy efficiency and EE programs is vital to implementation. Funding increased significantly in the early 1980s as EE and DSM programs in some parts of the country were incorporated into the utility's integrated resource management (IRM) plan. IRM is where a utility forecasts future energy demand and works to meet demand through a cost-effective portfolio of supply and EE and DSM options. This changed in the mid-1990s as state utility restructuring policies reduced or eliminated funding for energy efficiency as a resource, with the hope that competitive market retail power offerings would replace the need for policy-driven energy efficiency investments. As a result, ratepayer funding for energy efficiency decreased across the country (fig. 3–1).

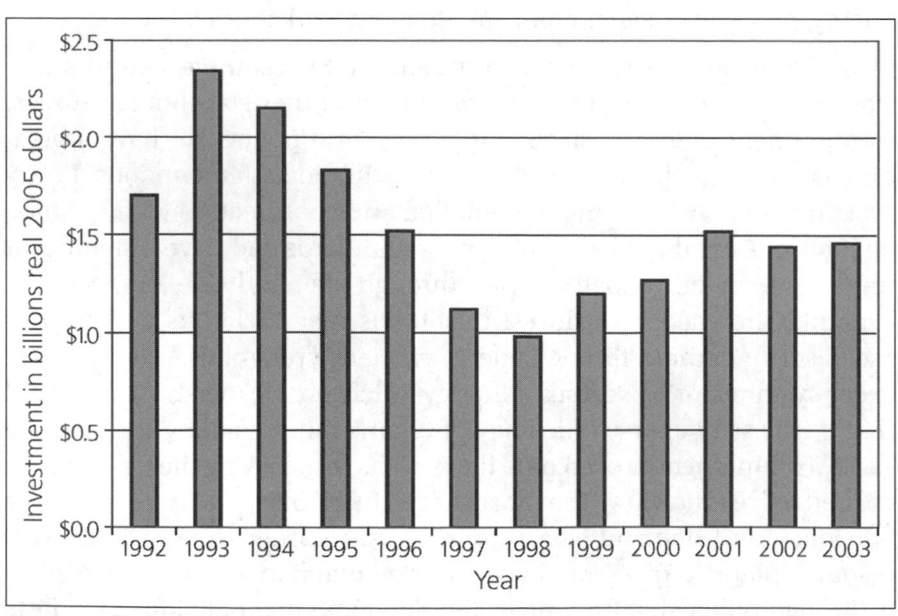

Fig. 3–1. Energy efficiency spending decline (Source: ACEEE 2005 Scorecard [York and Kushler, 2005; see also National Action Plan for Energy Efficiency, 2006, p. 1-5]; data adjusted for inflation using the U.S. Department of Labor, Bureau of Labor Statistics, Inflation Calculator)

Interestingly, given today's current environment, investment in energy efficiency is on the rise. The Consortium for Energy Efficiency notes that budgets for energy efficiency in 2007 totaled $3.7 billion and are rapidly increasing. In 2007, electric utilities increased budgets by 14%, while gas utilities increased budgets by 68% from the previous year. By all indications, the trend to increasing investment in energy efficiency will continue for many years.[5]

Energy efficiency is a smart investment. In fact, the return achieved by many program administrators is two to five dollars in energy savings for every dollar invested. Recognizing that market and institutional barriers impede the timely market adoption of cost-effective demand-side resources, additional states are currently introducing or restarting discussion on policies to promote demand-side resources, recognizing this as a way to address both rapidly increasing energy costs and climate change issues. These policies and discussions include ideas on not only funding EE programs but also providing incentives for high levels of performance in delivering programs.

Stable and multiyear funding is necessary to establish and maintain effective energy efficiency programs. Multiyear funding allows for long-term planning and investment in programs that have high upfront costs and supports the ramp-up in infrastructure needed to delivery additional efficiency programs. Additionally, multiyear funding allows time for much-needed consumer awareness and education. Furthermore, it provides market stability, which is crucial to encourage and create the enabling vendor infrastructure needed to implement energy efficiency programs. States employ a variety of mechanisms to promote demand-side resources, including rate-based recovery mechanisms and resource procurement funding, each of which has benefits and drawbacks.

Rate-based recovery mechanism

Recovery in this form integrates program funding as an element of the overall cost of service approved by regulators for recovery in rates. In some cases, it takes the form of a specific tariff rider for energy efficiency. The utility is expected to account for all planned and actual costs. In most cases, to provide for dollar-for-dollar cost recovery, the utility is expected to account for all actual program costs and file for regulatory approval of periodic rate adjustments used to account for the difference between planned costs reflected in rates and actual costs.

A system benefits charge (SBC) is another form of rate-based recovery mechanism. The SBC is a charge on the customer's bill that collects funds for energy efficiency programs. Benefits of this type of approach are that it provides stable program funding and that it is transparent to the consumer. Many states have successfully achieved significant energy savings by establishing and maintaining an SBC as a funding mechanism.

A downside to the SBC is that it is not tied to energy resource planning or procurement. Rather, the typical policy rational for an SBC is to simply maintain the general energy, economic, and environmental benefits of energy efficiency programs. The charge, usually set as a cost per kWh, limits program funding to a cap adopted from the perspective of limiting short-term rate impacts, regardless of the long-term value of efficiency as a reliable resource cheaper than supply-side resources. Systems benefit–funded programs are administered by an energy efficiency team within the utility, by a state agency or authority, or by a third party.

Capitalizing energy efficiency costs

In this model, costs of energy efficiency programs are amortized. This model has been employed by Vermont, for example. The downside to this approach is that recovery is delayed and can be diminished in future rate cases. In fact, Vermont has discontinued capitalizing energy efficiency and expenses all costs. To be successful, the utilities need a policy that defines the allowable return and identifies which costs are eligible.

Resource procurement funding

This mechanism places demand-side procurement on par with supply-side procurement. In this model, regulators require utilities to consider energy efficiency as a resource and to spend dollars to procure energy efficiency resources, just as they would for generation resources. This spending is typically part of the utility's revenue requirement and might appear to the customer as part of the supply or fuel charge, explicitly or embedded.

Rate Structures

Utilities have an obligation to provide safe, reliable service. In some cases, this means providing adequate supplies to meet customer demand for energy and capacity (i.e., obligation to service), as well as the maintenance of a reliable infrastructure to deliver energy resources (i.e., obligation to connect). In states where utility restructuring assigned the risk and responsibility of providing supply resources to market-based providers, the primary focus of the regulated distribution company is to maintain and expand as necessary reliability transmission and distribution facilities. Traditional cost-of-service rate-making provides a rate of return for the utility investors' risk on capital investments and keeps Wall Street shareholders confident. This rate of return takes into account inflationary cost increases, infrastructure investment, and the associated increase of revenues owing to electric sales growth. This structure drives the utility to maximize reliability and safety while minimizing costs. Under the traditional kWh framework, energy efficiency programs result in decreased sales making it more difficult to recover fixed costs; therefore, it is not in a utility's best interest from a sales perspective to help customers reduce energy usage.

This leads to an inherent conflict between the incentive to increase sales and the existence of important state, regional, and national goals to increase end-use efficiency and minimize the environmental impacts of energy production and consumption. Aligning utility and public interest goals by disconnecting profits and fixed-cost recovery from sales volumes, ensuring program cost recovery, and rewarding shareholders can level the playing field to allow for a fair, economically based comparison between supply-side and demand-side resource alternatives and can yield a lower-cost, cleaner, and reliable energy system.[6] Common frameworks include decoupling, modified rate structures, fixed customer charges, and loss-based revenue (LBR).

Decoupling

The most commonly discussed rate structure is decoupling. Decoupling is a fair and efficient means to design utility rates. It separates the recovery of utility return on equity from volume. The symmetrical nature of decoupling prevents a utility from increasing its earnings by increasing delivered volume because additional charges collected in this event are refunded to the customer. Moreover, it does not shelter a utility from increased costs or guarantee that the utility will achieve the authorized rate of return. Decoupling is a revenue recovery mechanism, not a cost recovery mechanism. The objective of decoupling is to align actual revenues with revenue targets.

There is limited experience with decoupling on the electric side. In 1982, the California Public Utilities Commission first approved electric revenue decoupling. In California, utilities file a general rate case every three years, using a forecast test year. Between cases, regulators adjust authorized revenues through a mechanism that links revenues to growth in the consumer price index. This mechanism amortizes over- and under-collection over the next year, incorporating them into rates. This keeps the utility whole.

The details on how to decouple can be quite complicated and are open to debate. Experience implementing a decoupled rate structure is limited. This type of framework requires accurate forecast and frequent rate adjustments. Regulators evaluating how to implement decoupled rates must proceed with caution to consider the impact on ratepayers.

Modified rate structures

Rate structures can be modified to allow utilities to receive their revenue requirement with less linkage to sales volume. In this type of structure, revenues may be connected to the number of customers instead of sales, or an escalation formula may be used to forecast the fixed-cost revenue requirement over time.

Fixed customer charge

Another rate design option is to move more fixed costs into fixed customer charges. Most customers are familiar with this type of concept. For example, a local telephone company has a customer charge just to have access to the telephone system. The challenge with this type of rate structure is that as more costs become part of the fixed customer charge, there is less incentive for customers to reduce their usage. There is also an impact on low-use customers who may pay more than their fair share for connection to the system.

LBR

An LBR framework is yet another option. This option replaces revenue lost because of reduced sales associated with program implementation. This type of lost-revenue recovery makes a utility whole for their investment in energy-saving programs. This type of rate structure still maintains the throughput incentive and does not address losses as a result of other energy efficiency initiatives. For example, if a city creates a community challenge to reduce energy usage, the reductions would not be addressed in the LBR calculation. Only the reductions associated with utility programs are included in the LBR adjustment.

There are pros and cons to each rate design, and tradeoffs must be considered as new rates and structures are evaluated. These rate structures attempt to make the utility whole for losses associated with energy management programs. These rate structures need to complement incentives, to create the most positive environment for success in EE.

Incentives for energy efficiency

Given the present environment, where energy efficiency is such an excellent investment to mitigate energy growth and to address climate change goals, innovation and creativity are needed to advance the technologies and processes. The best way to motivate in a competitive marketplace is to find funding and provide incentive mechanisms that benefit customers, utilities, and shareholders. This can be achieved through a variety of incentive mechanisms that have been applied to energy efficiency.

For the aggressive acquisition of cost-effective demand-side resources, timely recovery of resource costs associated with program administration is critical. With regards to incentives, energy efficiency program administrators, whether utilities or third-party providers, perform better when incentives are applicable to high achievement of performance. Within utilities, a positive profit incentive based on the performance of demand-side resource programs enables the dedication of more intellectual capital to develop solutions, as opposed to a regulatory requirement associated only with a negative penalty. Performance incentives garner the interest of an organization and allow EE to compete along with distribution, transmission, and generation.

There are arguments that incentives are not necessary. Without incentives, EE efforts would simply be a regulatory requirement and would clearly be viewed differently by the utility. The innovation and intellectual capital that would be invested when energy efficiency is in competition with other utility investments would likely not be applied in the same nature. Incentives can be in the form of shared savings, performance incentives, and bonus rate of return.

Shared savings. This model allows utilities to receive revenue equal to a portion of the savings value produced by the energy efficiency programs.

Performance incentives. This model rewards utilities on the basis of performance against agreed performance goals. Often the incentives are on a prorated sliding scale—the better the performance, the better the incentives.

Bonus rate of return. This model is used in Nevada, where the utilities receive a rate of return 5% higher than authorized in rates of return for supply investments.

Program Administration

A variety of models are in place to administer EE programs. Utilities are well positioned and are often called on to design and deliver programs. In fact, research shows that consumers will mention utilities as a source of information for energy efficiency more often than any other medium, and 76.9% of consumers feel that it is important or extremely important for utilities to offer energy audits and conservation products and information.[7] Other program administration models include state agencies, municipal organizations, and third parties. Regardless of how the programs are administered, certain common foundations need to be in place to ensure effective program administration. These include establishment of clear goals, regulatory oversight, stakeholder participation, and access to customer information.

Clear goals

Clear goals that align with a state's overall comprehensive energy plan are needed for successful EE programs. The goals are a starting point for program and portfolio design life cycle. Through these pre-established objectives, program administrators can create a portfolio that maximizes the investment and achieves the highest potential performance. The goals allow assessment and communication of success. Evaluation tools and processes will also need to align with the stated goals to ensure accurate measurement of performance against goals.

Integration of EE goals into the process of resource planning ensures the incorporation of measures that are consistent with all state goals, including environmental, capacity, and energy savings goals. This integration aids in long-term planning for new generation or transmission or other future energy resource requirements.

Regulatory oversight

Regulatory oversight can ensure that goals are established to achieve meaningful results and that the utility efficiency or demand-side programs are well managed and cost-effective. Regulators, who have an administrative mandate to oversee efficiency programs, take seriously their role to ensure that the programs are delivering cost-effective solutions to ratepayers.

Regulators will be involved in reviewing and approving efficiency plans, such as budgets and benefit/cost ratios for the program year or years, if a multiyear plan is submitted. Regulators will also review and approve annual reports filed by utilities in which utilities true up spending and production to ensure that cost-effective programs are actually being delivered.

Stakeholder participation

Utility-administered efficiency and demand-side programs benefit from stakeholder participation. Particularly in the case of a public utility administering a systems benefit–type program, stakeholder groups can be numerous and are often actively engaged in the process to ensure that their constituencies are well served by the program.

In Massachusetts, one of the most active stakeholders represents the low-income population. This stakeholder group has been instrumental in securing and maximizing delivery of cost-effective energy efficiency benefits to some of the utilities' neediest customers. This group also supports a network of localized community agencies across the state to ensure that this population segment is served from a holistic benefits approach and, more important, that they do not get lost in the fold. Because of this group's concern that there is still a large population of these customers who are not being served or who for whatever reason may not be aware of the help available to them, they expanded their partnership with program administrators to develop a unique marketing campaign called *Energy Bucks*.

The purpose of the Energy Bucks campaign is to raise awareness of potential financial and energy efficiency benefits this population segment may be eligible for and to expand enrollment of this eligible population. Expanded enrollment also means expanded delivery of energy efficiency products and services program administrators can provide. With its catchy tagline and supported by a statewide media campaign during the coldest and most costly months, Energy Bucks has proven to be very successful in terms of creating awareness and, more important, attracting new participants. Not only are these customers able to reduce their energy costs, but a whole new population of consumers is now able to benefit from the health and safety improvements often attributed to a properly weatherized home. The Energy Bucks campaign is a great example of how low-income advocates and stakeholders in conjunction with program administrators can work together and share a common goal to advance the common good.

Access to customer data

Utilities have easy access to customer usage and usage patterns. Utilities use these data for target marketing, to develop new programs, to assist customers in analyzing their usage, and to measure the impact of energy conservation measures (ECMs).

Many EE programs are administered by utilities. Utilities are well positioned to deliver energy efficiency as they are the nexus of the natural partnerships among builders, inspectors, customers, energy service companies, stakeholders, and policy makers. Customers often look to their utilities first for advice on energy management. For example, when customers receive their utility bills—and, in particular, a high bill—their full attention falls on the cost of energy, and their first call may very well be to their utility to understand why their bills are so high. In the best case, utilities can provide one-stop shopping not only to answer their question but also to give customers information to better manage their energy usage going forward. Confidence and trust by customers in their regulated utility facilitates their participation in the energy efficiency programs. Utilities that provide energy management programs find that the programs provide a value-added touch point with customers and result in higher customer satisfaction.

Utility program offerings can be maximized when there is coordination among the utilities to offer common programs across a state or region. This coordination can also bring market leverage when several program administrators from utilities together request bids for services.

Centralized nonutility administration

Utilities are not the only entities that administer programs. Other organizations, such as municipal aggregators, statewide agencies, and third parties, also successfully administer efficiency programs. State-administered programs have a benefit of clear alignment with state resource goals.

Efficiency Vermont is an excellent example of a centralized statewide efficiency program administrator. To reduce costs and align efforts to maximize benefits, the state created an *efficiency utility* to implement a statewide energy efficiency program, instead of a patchwork of programs operated by the states' 22 electric utilities. In 2000, the state selected Vermont Energy Investment Corporation, through a competitive solicitation, to establish and operate Efficiency Vermont under a performance-based contract that includes incentives for meeting and exceeding goals. Through an agreement with the state's electric utilities, Efficiency Vermont maintains a current database of all customer accounts in the state to support marketing, delivery, and tracking of efficiency measures, projects, and services.

Now in its eighth year of operation, Efficiency Vermont has helped Vermonters to save more than $200 million on their electric bills. To date, half of all Vermont homes and businesses have participated in and directly benefited from Efficiency Vermont's technical assistance and financial incentives. The saved kWh cost just 38% of what utilities otherwise pay for comparable electricity supply on the wholesale market. Today, Vermont is able to meet more than two-thirds of its load growth through energy efficiency and is on a path to avoid all load growth in the near future.

Other examples of centralized administration are the Energy Trust of Oregon, a nonprofit established by the State of Oregon; the EnergySmart Program, administered by the New York State Energy Research and Development Authority; Efficiency Maine, operated by the Maine Public Utility Commission; and the New Jersey Clean Energy Programs, administered by the New Jersey Board of Public Utilities.

With third-party administration, a utility whose revenues are based on traditional cost-of-service structures is still financially encouraged to build load and increase sales. These load-building initiatives may be contrary to the goals and messages of the efficiency programs. Success with third-party administration requires strong cooperation between the utility and the efficiency organization.

Establishing and Enforcing Building and Appliance Codes

Codes and standards complement ratepayer-funded demand-side programs by locking in the market gains of programs that successfully build the availability and use of new technologies and best practices. This regulation frees up funding for voluntary programs to build the market for the next tier of cost-effective demand-side resources. Leading examples of this include the adoption of federal efficiency standards for high-efficiency refrigerators, clothes washers, and residential central and commercial unitary air-conditioning equipment. In many cases, these successful program efforts included state support for federal adoption of the Energy Star label to build market recognition for new high-efficiency products. Adoption of the Energy Star label also established an energy performance test procedure and technical specification necessary for the eventual adoption of state and federal appliance efficiency standards. In all of these cases, ratepayer-funded programs played a major role in building the competitive market for high-efficiency products. In California, utilities earn performance incentives by building the case for the successful state adoption of new appliance efficiency standards. Most recently, the Federal Energy Security & Independence Act of 2007 set new federal standards for premium motors and incandescent lamps, formalizing successful ratepayer-funded efficiency programs for these products. Figure 3–2 provides an example of the market adoption of high-efficiency clothes washers in New England that in 2004 led to a federal efficiency standard, which became effective in 2007.

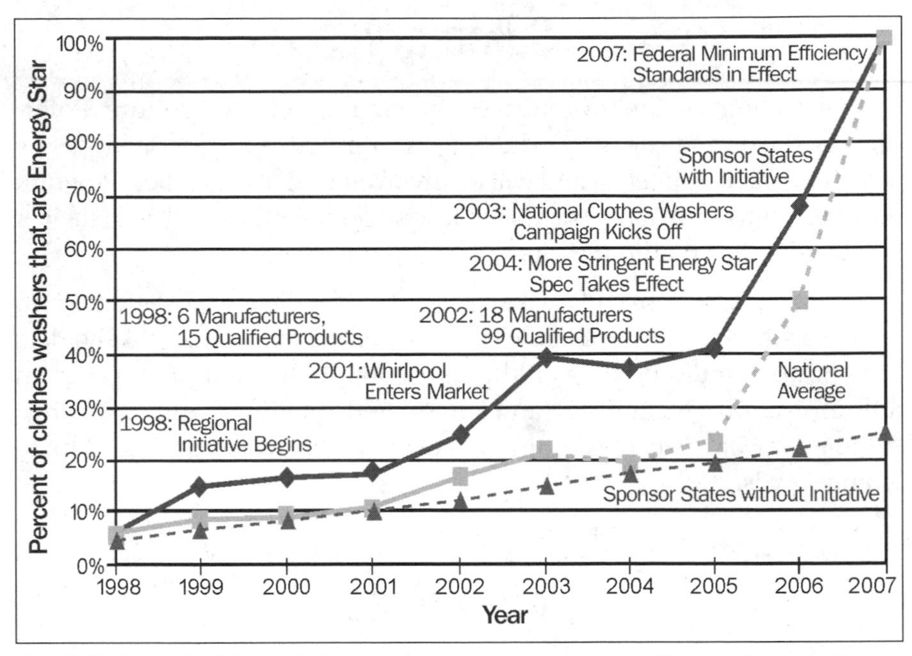

Fig. 3–2. Example of impact of programs and standards on clothes washer adoption (Source: NEEP 2006)

Another impact area from a policy perspective is the establishment and enforcement of building and appliance codes. Often codes require enactment of legislation. Program administrators can enhance building-code compliance with targeted programs on education and inspection. An ultimate target of addressing building and appliance codes is to achieve net-zero-energy buildings.

Code changes require coordination across multiple states and regions. Additionally, partnership and cooperation with manufacturers and retailers is essential to ensure that new codes can be manufactured and made available to consumers. This is where regional organizations like Northeast Energy Efficiency Partnerships can play a significant role in advancing policy outreach with respect to code adoption. Furthermore, regional and national organizations may aid program administrators in developing programs that go "above code."

Evaluation of the impact of code establishment and enforcement is challenging. Program administrators must understand the changes in codes and standards and the baseline against which future program impacts are measured.

Summary

Policy supports and even drives the creation of a EE culture. Policy creates the frameworks by which EE is funded and operated. Many stakeholders have interest and will be involved in shaping policy. Program administrators have a unique role to advise on policy, as well as to deliver on stated policy requirements.

Policy clearly is playing a key role in advancing EE programs and achieving market transformation. Utilities, policy makers, and stakeholders are engaged in discussions on the best combinations of rates, program administration, incentives, and mandated building and appliance standards to have the maximum impact on achieving energy and climate change goals.

References

1 Blume, Eric R. Powering up for America's future. *Electric Perspective Magazine.* Nov./Dec. 2007; vol. 32 No. 6. p. 44.

2 Vine, Edward. 2007. The integration of energy efficiency, renewable energy, demand response and climate change: Challenges and opportunities for evaluators and planners. Presented at the Energy Program Evaluation Conference, Chicago.

3 National Action Plan for Energy Efficiency. 2006. P. 1-5.

4 Coakley, Sue. 2007. Building energy efficiency programs in the Northeast. Presentation. Presentation to NY Public Service Commission; Case #07-M-0548 Energy Efficiency Portfolio Standard Overview Forum. July 19, 2007. Albany, NY.

5 Consortium for Energy Efficiency. 2008. Energy efficiency programs. In CEE 2007 report, p. 7.

6 National Action Plan for Energy Efficiency. 2006. P. 2-1.

7 Shelton, Suzanne, President & CEO, The Shelton Group. Shelton Group 2007. What are they thinking: Key insights on the consumer mindset regarding energy efficiency. Presentation. Nexus Client Conference March 2007, Scottsdale, AZ.

Part Two:
Deliver EE to Consumers

Chapter 4

Market Barriers and Assessment

Program managers offering energy efficiency, demand response, or distributed generation must build the program by starting with a solid understanding of the market they are trying to serve. This knowledge can be obtained through a market assessment, which provides program managers with information on the size, energy-use patterns, and preferences of the customer base they are serving.

Residential and commercial and industrial (C&I) markets have different attributes and challenges but share the common market barriers of awareness, availability, accessibility, and affordability. With both residential and C&I markets, EE program managers need to invest in research that provides insight into the size, characteristics, and potential of the market. A market assessment will also attempt to determine the viability of achieving the savings potential by gathering information on the barriers to adoption of energy efficiency.

Residential and small commercial markets, while different, lend themselves to market assessment using statistical sampling. Assessing large C&I customers is exponentially more complex and expensive than assessing residential and small to medium C&I customers because of the unique business processes, number of configurations of HVAC systems, and variation in the building envelope. A more customized approach to identifying and assessing large C&I opportunities is needed. Program managers will also want to tap into detailed usage information maintained on large C&I accounts as, generally, these accounts have more sophisticated metering, such as time-of-use pricing. In some organizations, the large accounts are assigned account managers. Tapping into the knowledge of the account manager is another tool for assessing the large C&I market.

Both residential and C&I program managers must build an understanding of their market or carve out a niche within the market, to design programs that are attractive and to market EE programs in a cost-effective manner. In chapter 2 ("Understand the EE Life Cycle"), a model for designing programs was introduced (see fig. 2-1). Figure 4-1 revisits this model and shows that the core of program design is the market assessment. From that basis, program type, communication and marketing channels, and delivery channels are defined.

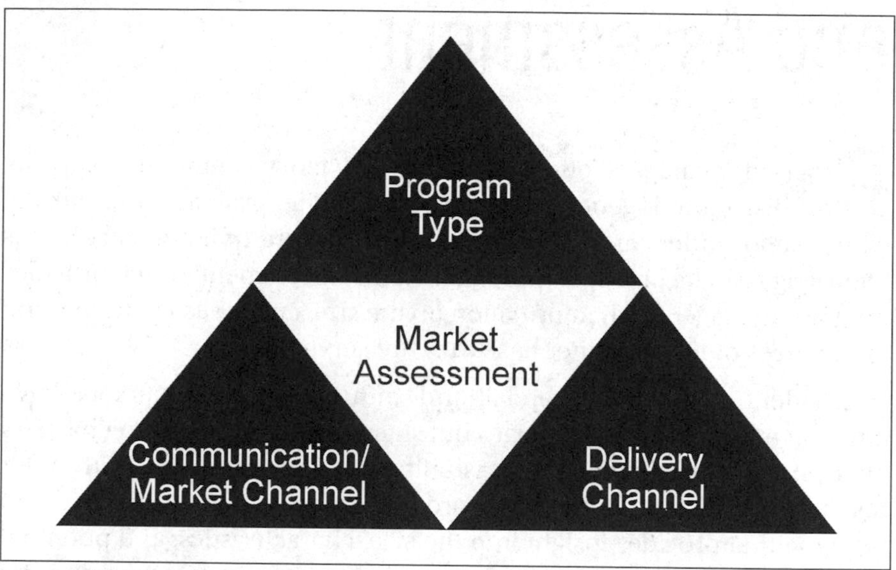

Fig. 4–1. Design process

In this chapter, a more thorough drill-down of how a program manager develops a strong understanding of the market is provided, featuring an overview of the residential and C&I markets and insights into the market barriers program managers may face when promoting EE programs. This chapter also offers a framework for both residential and C&I market assessment. This framework can then be used as a foundation for program managers to build successful programs around energy efficiency, demand response, and distributed generation.

Market Barriers

As with any market, there are barriers that must be overcome. The primary barriers to serving both residential and C&I markets fall into four basic categories: awareness, affordability, accessibility, and availability (table 4-1). All of these barriers may have an impact on consumers' adoption of energy efficiency and on the program delivery.

Table 4–1. Market barriers

Barrier	Description
Awareness	Lack of information on options
Availability	Lack of product manufactured
Accessibility	Lack of ability to access or obtain product
Affordability	Higher cost than alternatives

Awareness

The awareness barrier is encountered when customers lack information on available options. This barrier is largely dependent on communication channels. For the residential market, awareness is addressed primarily through mass-market communication channels, such as print, radio, and television. The C&I market is vastly different from the residential market as most awareness is derived from in-person sales calls with utility staff, communication via trade allies or industry associations, and targeted publication in trade journals. In both these markets, EE program managers recognize that, in addition to raising awareness among the end users, they must also raise awareness among key strategic partners, such as contractors and builders.

For residential programs, such as energy efficiency and demand response, which target the mass market, advertising is another important tool. To reach the residential market, program managers will leverage the traditional tools of bill inserts, bill messages, and the World Wide Web to raise customer awareness of energy efficiency programs. Program managers sometimes augment these tools with paid advertising through direct mail, radio, billboards, and, in some cases, even television.

Residential energy efficiency programs also benefit from the EPA's Energy Star program, advertising, and outreach. Through its established Energy Star program, the EPA plays a huge role in the residential market by raising consumer awareness of product and appliance energy usage on a national level. Energy Star provides consumers with an unbiased tool for comparing energy efficiency ratings of appliances. The Energy Star brand is often used as the foundation of individual administrator and regional marketing campaigns.

For C&I accounts, personal outreach is key. Utility key account managers, EE program managers, and direct sales staff play an important role in raising awareness within the C&I marketplace. Often these folks have built strong relationships with large C&I customers and have a strong understanding of their customers' businesses and needs. Personal outreach is made effective by arming key account managers, EE program managers, and direct sales staff with information and program collateral on programs and offerings. Supplementing personal outreach with targeted advertising in trade magazines or business journals is also a common tool within the C&I market.

For all EE and DSM programs and target audiences, outreach to builders, developers, equipment contractors, and other trade allies is an effective marketing tool. The time invested with these trade allies can be considerable but is often well spent. These trade allies can encourage consumers to consider the energy savings aspect of their purchase. They can also help consumers understand that the additional upfront costs associated with more efficient equipment are mitigated by reduced future operational costs. Program managers may tap into industry trade events, trade ally training events, and various home shows to reach this important market actor. In addition, program managers may proactively offer training and educational services for trade allies to increase awareness of programs. Finally, the Web is becoming an important tool to provide trade allies with information regarding programs; some program managers are offering trade allies custom-designed portals containing information that is pertinent to this important group.

Availability

The availability barrier arises when manufacturers either do not produce significant quantities or do not effectively market the energy efficient products they do make. Product availability is based on several factors, including market size, market demand, production capacity at the manufacturing level, and wholesale or retail inventory levels. Unlike other retail consumer merchandise, EE products or services often require longer market adoption periods and may be slower to achieve mass-market demand and capacity.

Incentives are used by program managers at all levels of the product delivery chain to increase availability to both residential and C&I markets. Incentives may be offered to manufacturers to encourage new product development or may be targeted to retailers and trade allies to promote adequate supplies and prominent shelf space of energy efficiency products. Incentives are often established by meeting or exceeding industry association guidelines or building codes. However, the lack of availability of products in such categories can impair the ability of customers to implement energy efficient equipment and earn incentives. Industry associations such as the Consortium for Energy Efficiency host meetings of program administrators and industry partners to explore new ways to develop or produce new alternatives to meet projected demand.

An example of incentives used to encourage product development is the Lighting for Tomorrow lighting-fixture design competition, sponsored by the Consortium for Energy Efficiency, the American Lighting Association, and the U.S. Department of Energy. In this case, incentives are used to encourage manufacturers to create new efficient lighting products. Winners of this annual competition are rewarded not only financially but also, more important, with product orders from big-box stores and specialty lighting showrooms.

Accessibility

Customers' access to (or ability to obtain) a more efficient product relies on distribution retailers to stock and display products in sufficient quantities and in visible areas so that consumers can find them. For both residential and C&I markets, addressing accessibility issues involves working with distributors and retailers to stock product. It also involves establishing a market characterized by contractors or installers with sufficient technical skills, experience, and certifications to successfully promote, sell, and properly install energy management solutions. For demand response, accessibility involves building a business case that leverages market rules and regulatory treatment to provide customer's access to demand response as an option.

One of the best ways to address accessibility in the mass market is to work with retailers to promote product on the floor. CFLs are a great example of how program managers can work with retailers to increase product availability. In the mid-1990s, while helping to mature this market, program administrators worked with retailers to establish a retail coupon model. This model was extremely successful early on, but as it grew, it became less and less cost-effective owing to the coupon-processing costs. An upstream buy-down model at the manufacturer's level was then developed to increase cost-effectiveness. In this model, manufacturers compete for program administrator dollars through negotiated cooperative promotions and in turn provide products on a mass scale at a discounted price to distributors and retailers. The savings are then passed on to consumers at the retail level, thus eliminating individual coupon-processing barriers and expenses.

In the commercial market, accessibility is most often achieved through energy services contractors. These contractors are often rewarded on a performance-based model in which they market directly to businesses, offer an array of new products and services, and in turn receive compensation from the utility for delivered savings. This type of model is extremely effective in getting the products to the market and providing a turnkey solution for business managers responsible for energy management and cost-effective capital improvements.

For demand response, accessibility may be a barrier to consumers because program managers have not been able to build a business case to offer demand response. Market rules or regulatory support may not be supportive of investment in demand response. In some territories, there is a disconnect between wholesale and retail electricity markets, sometimes referred to as a fractured value chain. In restructured markets, like Massachusetts, where the distribution utilities divested of generation, the fractured value chain is resulting in limited penetration of demand response on the mass-market level. The traditional business case of deferred generation cannot be built easily in this model. Often the regulatory perspective is not cohesive from a demand response perspective. Utilities may be allowed to get cost recovery for the investment but may not be allowed to earn a rate of return.

Building compelling business cases today for demand response must go beyond the benefit of deferring generation. Successful business cases include the benefits from a transmission, distribution, substation, and environmental perspective. Successful utilities and energy service companies do their homework in preparing a business case. Many will leverage providers in the marketplace, like Comverge, to help with identifying components and benefits of a business case. To build a compelling business case, it is helpful to reference the positive results from other implementations and consider using an outside consultant trained specifically in building demand response business cases.

Affordability

Higher cost associated with energy efficiency solutions, as compared with other, non-energy efficient alternatives, represents a final market barrier. This is because the more efficient products—being less developed—have higher first costs and have a smaller market share.

For both residential and C&I markets, rebates or incentives are one of the most effective and popular tools used by program managers. These direct consumer rebates or incentives are one of several financial tools that program managers can use to address affordability. Other tools include upstream rebates directed to retailers and distributors along with low-interest loans for home improvement projects.

Program managers must determine the appropriate amount of incentive or rebate to encourage the purchase of the more expensive efficient product. Often a program manager will start with an incentive equal to the incremental costs associated with the more efficient product and assess, in the evaluation process, whether the incentive is adequate or even needed.

With the C&I market, customers are heavily focused on the business case for energy efficiency to provide a competitive advantage. Energy efficiency programs compete for funding against other investment opportunities, such as labor productivity or investment in technology to improve operations. Supporting customers by providing financial data for EE projects is important. Too often, energy efficiency projects are presented in simple payback terms. For the C&I market, providing financial data on the return on investment or internal rate of return for EE projects helps to make these projects competitive with other capital expenditure options.

Affordability becomes a barrier for demand response because of the complexity and cost of the technology infrastructure to support the program. Customers are often charged a premium to participate in real-time pricing and demand response programs. This premium is associated with the large cost of offering this service to customers and is largely driven by increased metering, communication infrastructure, and billing system enhancements required in order to provide this customer option. Customers must achieve greater energy savings than the increased costs associated with participating in a real-time pricing program.

To conclude this discussion of market barriers, let us consider the success story of Efficiency New Brunswick (ENB). By addressing multiple barriers, New Brunswick has achieved one of the largest market shares of High Performance T8 (HPT8) lighting product of any jurisdiction in North America.

In 2005, the Province of New Brunswick established ENB as a crown corporation with the mission to help citizens and businesses in the province use energy more efficiently. Charged with quickly building new programs with a small staff, ENB looked for program options that could be implemented by the market with relatively little staff support and that could spread benefits broadly across the province. ENB soon identified such an opportunity with HPT8 lamps and ballasts. While HPT8s had been introduced into the U.S. market several years before, they were unknown and unavailable at New Brunswick lighting distributors.

Working with major lamp and ballast manufacturers and the province's electrical supply houses, ENB developed the Bright Ideas program. Bright Ideas eliminates the entire layer of consumer rebate application completion, processing, and verification by applying incentives directly to electrical distributors for sales of qualifying equipment. On a monthly basis, electrical distributors submit sales totals of qualifying equipment to ENB and receive reimbursement. Distributors run a sales report detailing the qualifying equipment and sales information, such as the shipped-to address for larger purchases. This enables ENB to determine the location of the installations and inspect them as necessary to assist with measurement and verification. Incentives were purposely designed to offset 100% of the incremental cost of qualifying products in most situations, thereby effectively eliminating the first cost barrier entirely. In addition to a product incentive, distributors are offered a small transaction incentive to offset their administrative and reporting costs.

As a result of this program, most of the supply houses converted their stock entirely to HPT8 products in less than six months, and it has been suggested that New Brunswick now has the largest market share of HPT8 products of any jurisdiction in North America.[1]

C&I Market Assessment

Successful C&I program managers understand their market from the perspectives of customer profile, business characteristic, and end use. Program managers will want to understand end-use saturations, energy consumed by end use, and even hourly load profiles for commercial segments. To gather this information, program managers will integrate on-site survey data with energy usage data.

Gathering commercial data is complex and expensive because of the number of different footprints, sizes, and operating schedules in the commercial market. Because the studies are so complex, they are often not completed or are out of date. In these situations, program managers must attempt to formulate an understanding of their market by using other means. Tapping into the Standard Industrial Classification (SIC) codes that are often maintained on customer information databases allows utility program administrators to segment the C&I market by business type. If these SIC codes have been well maintained, they can augment customer research and can provide managers with a good understanding on the market and associated energy efficiency opportunities. Customer research can add richness to the SIC codes and usage data by providing further understanding and data about building type, business characteristics, and end uses.

The industrial market represents some of the largest users, but energy demand varies across regions on the basis of the level and mix of economic activity, technology development, and raw materials. Generally, information gathered on the industrial sector will not be assembled through formal studies; rather, it represents a combination of data available from the customer information system, with knowledge from strategic account managers and others who are familiar with the business customers in the territory. C&I program managers will attempt, either with the benefit of formal studies or by use of less statistically based methods, to understand the attributes associated with the businesses represented in their territory, including the profiles, characteristics, and end uses.

Business profile

Commercial research will attempt to characterize C&I segments based on premises-level information. One benefit the C&I program manager has over the residential program manager is that the customer database already has business customers segmented using the SIC codes and includes related information on rate and usage history. Augmenting this knowledge with research helps to identify trends and characteristics associated with specific segments. Table 4-2 outlines common commercial segments.

Table 4–2. Common commercial segments and definitions
(Source: U.S. Department of Energy, Energy Information Administration [http://www.eia.doe.gov/emeu/cbecs/building_types.html])

Commercial and Industrial Segments	Definition
Education	Buildings used for academic or technical classroom instruction
Food sales and service	Buildings used for retail or wholesale food or for preparation and sale of food and beverages for consumption
Health care	Buildings used as diagnostic and treatment facilities
Lodging	Buildings used to offer multiple accommodations for short-term or long-term residents
Manufacturing	All establishments engaged in the mechanical or chemical transformation of materials or substances into new products
Mercantile	Buildings used for sale or display of goods other than food
Office	Buildings used for general office space, professional or administration offices
Public assembly, order, religious worship	Buildings where people gather for social, recreational, preservation of law and order, or religious activities
Service	Buildings in which some type of service is provided other than food service or retail sale of goods
Warehouse and storage	Buildings used to store goods, manufactured products, merchandise, etc.

While the categories listed in table 4-2 are some of the most common, program managers may want to build greater understanding of subcategories within these building types. For example, data centers, which are intense electric users, may be a growing subsector within the category of office space in certain service territories.

A program manager must be engaged in defining exactly how detailed the data collections should be to provide the meaningful information that is needed to build programs. Additionally, if there are some unique market segments, a program manager may want to include these in the data collection effort to better understand usage. These data will help program managers get segment-level consumption.

Business characteristics

Business characteristics provide insight into fuel type used, building size, age, and envelope components among other attributes (table 4-3). The building characteristics help program managers understand the size and age of the existing building stock. Combined with other data attributes and profiles, these data can help to build targeted programs or approaches to address common issues.

Table 4–3. Building characteristics and definitions

Business Characteristic	Definition
Demographics	Information that describes customers' investment and interest in energy
Heating fuel and system	Information on heating systems and space heated
Cooling and ventilation	Information on cooling systems and space cooled
Operating schedules	Understanding of the times of operation and frequency of change
Building envelope	Information on the envelope separating conditioned space from the outdoors
Load profiles	Definition of a customer's usage by time
Backup generation	Additional generation required to support continuous business operation

Demographics. In this category, information on customers is gathered to understand their interest and investment in energy. What percentage of a given customer's business budget is targeted to energy? How is the business impacted by power quality?

Research in this area may also include gathering data on participation in energy efficiency programs, the drivers for investments in energy, and the percentage energy represents in the annual operating costs of certain business types. Interest in renewable or highly efficient distributed generation or interest and ability to reduce load during peak periods be may also be investigated. Research may also probe interest in programs and services, such as demand response, reliability programs, and energy management programs. Additionally, what is the dominant electric and gas rate used by the facility? In large territories, understanding the climate zones may help to calibrate the data. Other data that may be helpful are the number of full- and part-time employees who occupy the facility and whether the facility leased or owned.

Heating fuel and system. Program managers will need data on the heating systems and space heated. For example, what fuel is used—electricity, gas, or oil? Do they have forced-air systems or hydraulic? Are these systems roof mounted? What is their age? Is this the main heating system, or are there other heating systems? What percentage of the floor space is heated by the system? What is the square footage of space heated? What are the replacement plans for the system? Other heating options to be surveyed include wall or ceiling units or baseboard units. For these units, it is also necessary to understand whether they are main or supplemental systems.

Cooling and ventilation. Understanding the cooling and ventilation systems, along with the space cooled, is also important. Is the ventilation provided by a central system, local fans, or open windows? It may also be necessary to determine the square footage of space that is cooled and obtain a deeper understanding of the types of cooling fuel and systems. Similar to heating, determining whether the air conditioning systems are roof mounted and whether they are main or supplemental systems is beneficial. The facility may also use wall or window units, swamp coolers, or chillers. The use, age, and square footage cooled are other specific data points that provide insight into usage profiles.

Operating schedules. Because of the complexities of the C&I market, it is important to drill down one more level to understand the various operating schedules and the frequency with which they change. Do the schedules change daily, weekly, or monthly? What are the heating hours and the cooling hours? When is the facility open and closed, by day of the week? What are the business-hour heating settings versus non–business-hour heat settings? Similarly, what are the cooling settings? Does the facility have programmable thermostats to support their operating schedules? What is the load factor for the facility? On a more sophisticated level, does the facility use an energy management system to manage not only heating and cooling but also other equipment? For example, within the restaurant segment of the food service industry, operating schedules may be more skewed to midday through early evening. By contrast, convenience stores may operate on a seven-day-a-week, 24-hours-a-day schedule.

Building envelope. The commercial building envelope includes the insulation, windows, doors walls, foundation, floor, ceiling, and roof and is the barrier between the conditioned indoor environment and the outdoors. The building shell can have a large impact on the effectiveness and associated cost of heating and cooling. Building-shell information will provide details about the shape and size of each building, the wall and roof coverings, window types installed, number of floors both above- and below-ground, building orientation, and building materials used in construction. The more data collected about these building characteristics, the more accurate the analysis will be. Program managers considering programs that offer insulation services need to know the existing levels of wall, ceiling, or attic insulation to best determine realized fuel savings. In the C&I market, programs on window shading, for example, may be offered to reduce heat gain.

Load profiles. Load profiles define a customer's energy usage versus time of usage. These load profiles are essential to energy analyses and graphically show the variation in a customer's electrical load over time and are used to identify opportunities. When properly displayed, they can graphically show anomalies and areas of intense usage on which to focus. They are used to identify areas of energy efficiency opportunities, investment presentations, and energy tracking. Typically, the utility has these data available for customers from its load research group who have advanced meters. Providing these data helps in analyzing the usage behaviors by time of day, day of week, and week of the year. It helps in analyzing coincident peak and non-coincident peak for the commercial segments.

Backup generation. For demand response or distributed-generation programs, information on customers' need or investment in backup generation to support continuous operation is important. Are they interested in renewable generation? Do customers have backup generation or need backup generation from an operational standpoint? If generation is on-site, what is the size, type, and age of the system? What load or business operations does this system support?

Program manager will need these data to determine a target market. For example, customers with backup generation may be a highly attractive target market for demand response programs. Customers with older distributed-generation installations or with interest in renewable energy are targets for distributed-generation programs.

End uses

Data in this final category help program managers to gain an understanding of the end uses by type of facility. End uses of interest in the commercial market include space heating, cooling, and ventilation; lighting; water heating; business equipment; and refrigeration and cooking. Figure 4–2 shows the relative energy use by C&I end-use categories. Clearly, HVAC represents the largest opportunity.

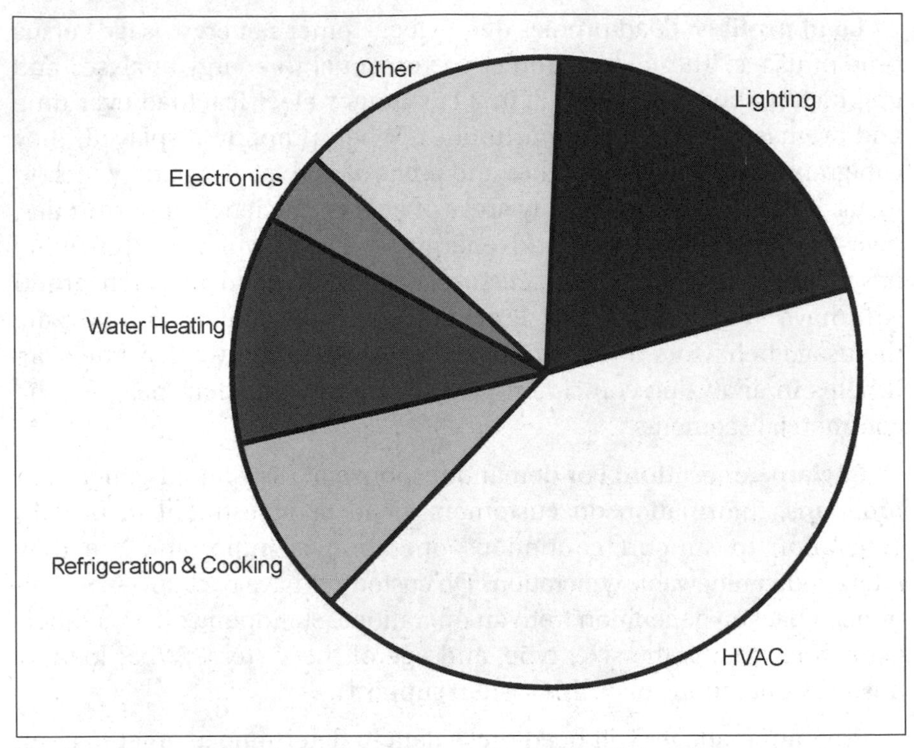

Fig. 4–2. Commercial end-use splits (Source: U.S. Department of Energy [*Buildings Energy Data Book*, 2004, Table 1.3.3])

Water heating. Water heating is a large end use in commercial buildings, representing up to 15% of the commercial-sector fuel use.[2] Researchers will attempt to understand the fuel used for water heating, as well as the number of water heating systems, their age, and size of the systems. Program managers will use this information to identify retrofit opportunities. From a demand response perspective, water heating provides an opportunity for interruption during peak periods.

Business equipment. Although this can be a catchall category, it represents significant commercial use, at approximately 7%.[3] Business equipment includes items such as personal computers, printers, fax machines, copiers, servers, flat-panel televisions, and local area networks. An inventory of the number of pieces of equipment, by type and the percentage of each type that are Energy Star compliant, is valuable.

Lighting. Lighting in the commercial market is a large end use and, for energy efficiency program managers, offers some of the most cost-effective programs opportunities for obtaining energy savings. This is accomplished by replacing inefficient equipment with higher-efficiency lighting. Programs managers will want to understand the number and type of outdoor lighting fixtures and the hours of operation. As with indoor fixtures, what are the types, the numbers, and hours of operation?

Refrigeration and cooking. Refrigeration and cooking can be a large energy user and is often driven by the customer segment. The food service industry will require more refrigeration than other types of businesses. With refrigeration, understanding the commercial types of refrigerators and freezers is important. Are they front opening with a door, retail display, or walk in? How many are there of each type, and how they are fueled? For cooking, program managers need to know whether their customers are using commercial ranges ovens, mixers, standard refrigerators, and freezers. They may also need to know if their customers use dishwashers and dishwasher boosters. Program managers will often need these data organized by type of business within a certain sector. For example, the aforementioned data might need to be reported separately for restaurants and for convenience stores.

Motors. Motors are a common tool in industry and industrial processes. Efficiency gains can be realized from utilizing high-efficiency motors; however, considerable opportunities exist in managing the controls and processes involving the motor. Many systems are complex and offer opportunities for redesign to downsize, modulate, or even eliminate motors.

Residential Market Assessment

Successful residential program managers fully understand their residential market not only from a perspective of customer profile but also from a housing and end-use perspective. Table 4–4 outlines typical attributes important to understanding the residential market. Program managers gather attribute data and build their market understanding on the basis of consumer research. This research can be conducted using market surveys that provide statistically based data on each of these attributes in the program manager's service territory. This section will review these attributes and discuss the data program managers need to secure in order to cost-effectively understand and serve their market.

Table 4–4. Framework for residential market assessment

Customer Profile	Housing Characteristics	End Uses
Interest in services	Fuel types	Appliances
Spending priorities	Size and age	Water heating
Demographics	Building shell	Home electronics
Behavior pattern	Heating and cooling practices	Lighting

Customer profile

Obtaining and using an understanding of the consumer is vital for successful programs. For successful design and implementation of EE programs, it is important to know the interest level of consumers in various products and services, along with their spending priorities, demographic information, and behavioral patterns.

Interest in services. Program managers need to understand the various services consumers would like their utility to provide. Consumers may be interested in green power, budget plans, fixed price options, time-of-use rates, automatic payment options, electronic billing, energy management solutions, wiring protection plans, surge protection, heating protection plans, and Web self-service, among other services. Knowing which services consumers are interested in allows program managers to link an electronic energy management products and delivery channels to interest in those services. For example, consumers interested in electronic billing may be good candidates for an electronic energy management tips newsletter. Consumers interested in green power may be more likely to invest in energy efficiency solutions to help the environment. Program managers will also want to understand interest in these services by customer segment, by income level, by fuel type, and so forth.

Spending priorities. Spending priorities reflect how customers believe utility energy efficiency dollars would be best invested. For example, program managers will find value in understanding how consumers value rebates and loan products and whether they value home energy audits over more generalized educational materials or communications. In this area, research consistently indicates consumers prefer rebates. Program managers may want to conduct further research to better understand preferences with respect to rebates, such as whether rebates for heating systems are more popular than rebates for windows or other types of appliances. Program managers will also want need research on these attributes by customer segment.

Demographics. Program managers need to understand their consumer demographics from the perspectives of income, age, and ethnicity, among others. These becomes important data to cross-tabulate to other attributes, such as spending priorities and housing characteristics. One key element from an income perspective is to understand the low-income market segment and, within this segment, spending priorities, type of housing, and interest in utility services. This becomes vital information to a program manager designing a suite of low-income programs.

Behavior pattern. Because the residential market is large and diverse, program managers may find it helpful to understand variances in consumer behavior patterns. Consumer research can be conducted to help program managers to better define market segments by behavior patterns. This information can be helpful in targeting the market and conducting outreach activities more cost-effectively.

- One model for this was developed by IBM Institute for Business Value. This model identifies four segments among residential and small commercial electric power consumers: *Passive ratepayers* are consumers who are relatively uninvolved in energy usage decisions and are uninterested in taking responsibility for these decisions.
- *Frugal goal seekers* are consumers who take modest actions to address energy usage but are constrained in what they are able to do because of their disposable income. Low-income customers may fall into this category.
- *Energy epicures* are high-usage consumers who have little or no desire to practice energy conservation.
- *Energy stalwarts* are consumers who have specific goals or needs related to energy usage and have the income and desire to act on these goals.

Interestingly, the data from IBM suggest that almost 50% of the consumers in the United States are passive ratepayers.[4]

Housing characteristics

For residential programs, the opportunities for large energy savings in energy through efficiency programs center on the house. Space heating and cooling represent over half of the residential end use and hence offer some of the most significant opportunities for energy savings in the residential market. Therefore, it is very important for program managers to understand the building stock within their service territory. Program managers may include additional categories of research on housing depending on the service territory. Warm climates, for example, may want to gather data on residential pool operating characteristics. This knowledge will support program design.

Fuel types. Information on the fuel type used for heating, cooking, and water heating is the first key data point. What percentage of consumers use natural gas or have access to natural gas or an alternate fuel? What percentage of consumers use electricity for water heating or oil for water heating? What percentage of these consumers could switch to a more efficient fuel type?

Of particular interest to a program manager, for example, would be how many customers have access to natural gas but use electricity for water heating. These customers become a prime target market for promoting high-efficiency heating and water-heating equipment, as well as a more efficient fuel type.

Size and age. Understanding the housing stock from a size and age perspective is also important. What is the percentage of homes built before 1960, between 1960 and 1990, or after 1990? What is the size of the home in square feet? How many rooms do the homes have?

This information aids program managers in determining the most viable target markets for services. Older homes are more likely to benefit from energy-saving measures than new homes. Many old homes were not built with wall, floor, or attic insulation. Older homes are likely to have older windows, resulting in substantial heat loss. They are also more prone to have developed air leaks over time. Program managers will compare information on consumer housing characteristics as they define programs and determine the program potential.

Building shell. The residential home envelope includes the insulation, windows, doors, walls, foundation, floor, ceiling, and roof and is the barrier between the conditioned indoor environment and the outdoors. The building shell has a large impact on the effectiveness and associated cost of heating and cooling. Program managers considering programs that offer insulation services need to know by fuel type whether homes have wall, ceiling, or attic insulation.

Window rebate programs are often considered by program managers. In this case, managers need to understand the penetration of double-pane versus single-pane windows, as single-pane replacements provide the greatest and most cost-effective energy savings opportunities.

Sealing the residential home against air leakage is another savings opportunity. Air leakage can be through gaps between the framing materials and improperly installed insulation, through holes drilled for plumbing and wiring, and around doors and windows. Program managers may offer consumers a blower door test to check for adequate sealing or to identify where additional sealing can be completed.

Heating and cooling practices. Nearly 30% of a typical home's utility bill goes toward heating and cooling.[5] Energy efficient building envelope and properly sized HVAC systems can reduce this percentage. Program managers will be interested in the current penetration of heating and cooling systems. For example, what percentage of customers have room air conditioners? How many rooms on average have a room air conditioner? What percentage of customers has central air-conditioning? How old are the main heating systems, and which fuel are these systems supported by—electric, natural gas, or oil? Do the consumers use programmable thermostats or regularly set back manual thermostats? All of this information constitutes a powerful asset for program managers designing programs to serve their particular customer base.

For example, data indicate that in 25% of homes the primary heating system is over 20 years old. In New England, though, this percentage increases to 30%. With the larger percentage of older heating systems, New England program managers have more opportunities for high-efficiency heating-system replacement.[6]

End uses

The largest uses of energy in the average U.S. household are space heating and cooling. From an electricity perspective only, the largest users of electricity in the household are appliances, including refrigerators and lights (fig. 4–3). Program managers must also be aware of the growing areas of opportunity for energy savings in home electronics, domestic hot water, consumer-based behaviors, and phantom loads. Program managers need to know what appliances are in the home. For example, what types of lighting are consumers using, and what is the penetration of the various types of lighting?

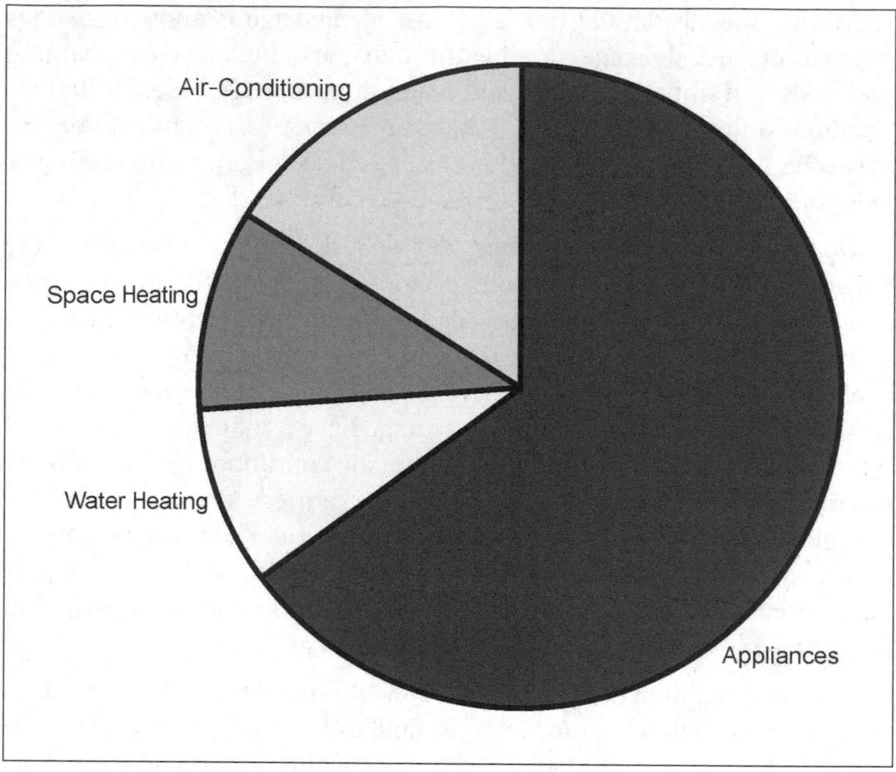

Fig. 4–3. Residential end-use splits
(Source: U.S. Department of Energy, Energy Information Administration [http://www.eia.doe.gov/emeu/recs/recs2001/enduse2001/enduse2001.html])

Appliances. Program managers will want to have an understanding of the percentage of customers that have dishwashers, clothes washers, and dryers. Specifically with regard to dryers, what percentage are electric versus gas dryers? How many customers have more than one refrigerator? Do they have stand-alone freezers? Do they maintain a second refrigerator?

In addition, program managers will want to know the characteristics of the appliances. How old are the appliances? What percentage of clothes washers are front-loading washers? Armed with these data, program managers can determine where rebates may be attractive in their target market. For example, several years ago, 88% of customers in the NSTAR territory were found to have top-loading washers, indicating a large opportunity to encourage customers to invest in the more efficient, front-loading washers.[7]

Water heating. Program managers will also want to understand the penetration and age of the water-heating systems in their territory. Do customers have electric or natural gas systems? What percentage of customers use oil for water heating?

Program managers may augment the research on end users with research on contractors', builders', and plumbers' awareness of and rates of installation of high-efficiency equipment. With water heaters, the replacement market owing to failure is one where speed of replacement is critical. Therefore, consumers will often select a replacement product on the basis of what a plumber stocks and has available. To identify and develop the best programs, program managers need to understand not only the needs of the end-user market but also the sales practices of the key trade allies in the market, such as plumbers and contractors.

Home electronics. The home electronics market includes computers, monitors, home servers, televisions, set-top boxes, digital video discs (DVDs), power supplies, printers, answering machines, copiers, video equipment, and stereos, among other devices. This is an area of explosive growth, along with trends of convergence, miniaturization, and transition from analog to digital technology. This industry, in partnership with government, has been successful in voluntary efforts to set standards and encourage innovation. From a consumer perspective, program managers need to understand not only the penetration of computers or televisions but also the size of the equipment—and, in the case of televisions, whether it is plasma screen, liquid crystal display (LCD), or other.

Another important aspect is consumer behavior. Are consumers leaving their electronics on and plugged in all the time, or do they turn them off when not in use?

Lighting. Lighting is a significant end use, representing over 16% of the energy usage in the residential home.[8] Lighting is delivered via fixtures, lightbulbs, and recessed downlight for residential applications. The efficient technologies underlying the future of fixtures and lightbulbs are fluorescent and solid-state lighting. The most common lightbulb is still the incandescent bulb, invented by Thomas Edison in the late 1870s. This mature, proven technology remains prevalent even though it is highly inefficient by today's standards; incandescent bulbs are inefficient because most of the energy they consume is used to produce heat, rather than light.

Program managers need to have an understanding of the penetration of different lighting technologies—specifically, CFLs in the residential marketplace. Additionally, managers may be interested in customer behavior around replacement bulbs. Do consumers replace lightbulbs with CFLs when a CFL fails, or do they switch back to incandescent?

Summary

A market assessment is the foundation to designing cost-effective programs to which consumers will respond. The assessment seeks to build a deep understanding of the customer base. This understanding supports program managers designing programs to reach the target audience.

For small C&I and residential customer segments, market assessment can be done in a survey form, using statistical modeling. The C&I assessment involves understanding the size and nature of the C&I segments and the associated business characteristics. Additionally, data will be gathered in the assessment of end-use categories and behaviors.

Residential assessments involve understanding the customer profile. Data are also gathered on end use and housing characteristics. Program managers may append the research with additional data, such as usage patterns and load profiles that are available in billing and meter data systems.

Program managers will use the market assessment data to design their programs. More information on EE program design—including types of programs, communication channels, and delivery channels—is provided in the upcoming chapters.

References

1 Bastion, Doug. principal North Atlantic Energy Advisors. Personal communication to Penni McLean-Conner, June 24, 2008.

2 Sezgen, Osman, and Jonathan G. Koomey. 1995. Technology Data Characterizing Water Heating in Commercial Buildings: Application to End-Use Forecasting. http://enduse.lbl.gov/Info/37398-abstract html

3 Ibid.

4 Valocchi, Michael, Allan Schurr, John Jullano, and Ekow Nelson. Plugging into the Consumer: Innovative Utility Business Models for the Future. IBM Global Business Services, p. 11.November 2007, Cambridge, MA.

5 U.S. Department of Energy. 2004. Buildings Energy Data Book. Washington, DC: U.S. Department of Energy, table 1.2.3.

6 KEMA Inc. 2004. Residential Market Research, vol. 1, Report of Findings: Residential Appliance Saturation Survey (RASS). KEMA Inc.

7 U.S. Department of Energy, Energy Information Administration. End-Use Consumption of Electricity 2001. http://www.eia.doe.gov/emeu/recs/recs2001/enduse2001/enduse2001.html

8 Ibid.

Residential Energy Eficiency

The residential market is large and diverse, representing a tremendous growth opportunity for energy efficiency. It constitutes 21% of the total energy usage in the United States, just below the transportation and industrial sectors in terms of total usage; most of this usage, over 60%, is to heat space and water.[1]

Utilities and other program administrators have been designing and delivering programs to the residential marketplace for over 30 years, successfully building on a deep understanding of the market and combining various program types, communication channels, and delivery channels (fig. 5-1). During this time, the technologies have evolved, and program delivery techniques have matured, resulting in significant energy savings for residential consumers. For example, the American Gas Association has documented that over 50% of energy savings comes from residential programs.[2]

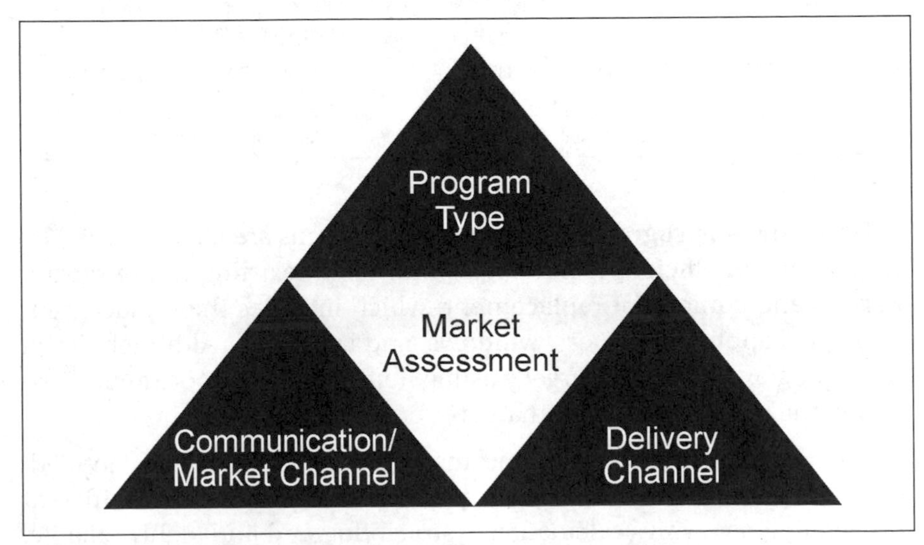

Fig. 5–1. Framework for program design

This chapter offers a framework for residential program design. Program managers will want to build on a market assessment (reviewed in chap. 4), with a program design framework to create programs best suited for their particular audience.

Residential Design

To successfully design programs, program managers must decide on the type of program, how the program will be communicated, and define the delivery channels. The combination of program type, communication and delivery channel will be designed to most cost-effectively target the desired audience.

Program types

Energy efficiency programs fall into three categories: direct install programs, education and audit programs, and rebate and loan programs (table 5–1). Program managers will use these types of programs individually or in combination as they develop their specific portfolio of offerings.

Table 5–1. Overview of program types

Program Types	Definition	Audience
Direct install	Installation of efficient technologies	Customers doing home retrofits or equipment replacement
Audits and education	Provide advice to consumers	End-use customers and trade allies
Rebates and loans	Financial incentive to encourage energy efficient investment	Customers doing retrofit or equipment replacement; trade allies

Direct install programs. Direct install programs are characterized by installations of efficient technologies and target existing-home retrofit markets and equipment replacement, which includes the replacement of lighting, appliances, doors, windows, and insulation. Although direct install programs primarily target existing home retrofit, opportunities also exist in the new-construction market.

The savings associated with direct install programs are easily quantified. For example, if an old refrigerator is replaced with a new efficient refrigerator, the savings calculation is straightforward and highly reliable, based on usage and technology characteristics. There is a high confidence level in the savings estimates for equipment replacement programs.

Audit and education programs. One of the most common energy efficiency programs in the residential market sector is the home energy audit. Home energy audits advise consumers on ways to lower energy costs, suggest energy efficiency retrofits that provide cost-effective payback on the investment, and often provide incentives to entice consumers to make such investments.

Program managers know that offering educational tools to consumers is powerful, raising awareness of and generating interest in programs. Web-based audits are now a common offering. These audits are designed to provide consumers with visual energy consumption portfolio by end use and make recommendations on available incentives and programs for which they may qualify. Another awareness-raising tool is to educate consumers by use of school-based programs targeted to youth or displays and event participation targeted to the community. Program managers may also target educational offerings to trade allies; examples of this include building code, quality installation practices, building science, and new-technology training.

Rebate and loan programs. Rebate and loan programs are a principal motivator for enticing consumers to invest in efficiency opportunities. Rebates are a powerful tool for a program manager. Data indicate that rebate programs have more participation than any other type of residential program offering.[3] Rebates are also directed upstream, to retailers and distributors. These upstream rebates address market barriers of availability and accessibility and provide a great mechanism to create partnerships with the retailers and distributors.

Program managers must determine the appropriate incentive or rebate amount that is needed to encourage the purchase of the more expensive efficient product. Often a program manager will start with an incentive equal to the incremental costs associated with the more efficient product and will later, in the evaluation process, assess whether the incentive is adequate, can be lowered, or can be eliminated altogether. These decisions are primarily based on market-share growth patterns and overall consumer acceptance of the product.

While less attractive to consumers, another incentive option offered by program managers is low-cost financing. In some cases, program managers will offer consumers a choice of a low- or zero-interest loans or a direct rebate. To offer loans, program managers must be able to either bill for the loan repayment directly or partner with a financial institution.

Communication channels

The residential market is large and diverse, with many communication channels. The challenge for a program manager is to identify which channels are the most appropriate to promote a particular program, understanding that each program has a cost that can vary from low to premium.

This section provides a channel management assessment tool for matching communication channels to programs (table 5-2). The tool highlights the typical costs per transaction or per customer, using the following categories:

- ***Low cost***—ranging from $0.00 to $3.00
- ***Moderate cost***—ranging from $3.00 to $10.00
- ***Premium cost***—greater than $10.00

Table 5–2. Communication channel assessment tool

Channel	Cost	Audience	Business Driver
Bill insert or message	Low	All customers	Promote awareness; targeted messaging ability
Direct mail	Low	Targeted audience	Targeted offerings to specific audience
Web	Low	Computer-savvy customers	Promote awareness, augment other channels
Paid print media (newspaper, bill board, etc.)	Moderate	All customers, but can be targeted message	Promote awareness; broader messaging visibility
Paid electronic media (radio, TV, etc.)	Premium	All customers, targeted message	Promote awareness; maximized messaging visibility
Community	Premium	Trade allies; dommunity events	Promote awareness; augment other channels
Personal outreach	Premium	Trade allies; targeted customers	Promote awareness; educate on programs

Bill insert and bill message. For utility program administrators, a very-low-cost channel is the bill insert and bill message. Data indicate that approximately 30% of customers will read a bill insert. A bill message, particularly if it is customized, can be very powerful. When customers are looking at their bills, energy costs are clearly on their minds. A well-targeted bill message or bill insert can very effectively drive program interest and augment other more expensive communication methods.

Direct mail. Direct mail is a low-cost, effective advertising tool that raises awareness and sparks interest and action in specific program offerings. Direct mail campaigns are typically consumer-based campaigns that enable administrators to target the market on a broad scale or by segment, based on certain demographics. Direct mail produces one of the highest response rates among the various advertising mechanisms, especially when limited dollars are available. The value of the direct mail offering is its ability to target a specific audience by use of a compelling message that drives action.

World Wide Web. The Web is a growing—and very cost-effective—communication channel for computer-savvy residential customers. As such, several educational tools are commercially available, along with custom-designed tools developed by program administrators. Program administrators use the Web primarily for education, augmentation of existing programs, promotion of programs, and lead generation. Online energy audits and appliance calculators are among the more popular tools that may be offered.

Online energy audits offer consumers the opportunity to analyze their energy usage. These audits provide real-time feedback to consumers on ways to save money and estimate energy savings.

The appliance calculator is another popular tool for the Web. These calculators are interactive and allow consumers to input information based on their specific appliance quantity and hours of use. In turn, consumers can see how much their devices are really costing them, and, with more advanced versions, consumers are able to compare their appliances or electronic equipment to similar Energy Star–rated products.

Program managers may also provide energy educational material on the Web. This material may be targeted to school-age children to expose them to the concept of energy conservation, as well as to energy savings ideas.

Paid media. For the residential market, paid media, both print and electronic, constitute a very important tool. This can be radio, television, billboards, and print advertisements in newspapers or magazines. This category also contains program collateral that can be distributed at events. Each of these channels comes with a premium cost, with television being at the high end of the scale. With paid media, frequency matters; hence, the budget to support paid media must be robust.

There are nontraditional venues, such as running advertisements in movie theaters before feature films. Print and broadcast media can be expensive, depending on the geographical market. Since advertising can be quite costly, program administrators may partner to develop a common campaign and, in doing so, stretch their limited marketing dollars.

An example of program administrators combining on outreach is the Massachusetts MassSAVE home energy assessment program, which delivers uniform consumer messages through statewide advertising campaigns. The value of paid media, either print or electronic, is the ability to reach large audiences with repetitive messaging, thus increasing the likelihood that consumers will identify with or react to the messaging or program offer.

Community. The local community is becoming an increasingly important communication channel to promote programs and increase participation. Sometimes called *community-based societal marketing,* this type of outreach is predominantly led by community leaders at community events. There is a powerful encouragement to "keep up with the Joneses" in this type of marketing. The cost for this outreach can be high because of the need for staff to attend events in the community. Community-based marketing will also include other communication channels, such as direct mail, program literature, and oral presentations targeted to customers within that community.

Increasingly, traditional community-based outreach is being augmented by viral marketing. Viral marketing comprises the marketing techniques that use preexisting social networks to create brand or campaign awareness. This marketing is grounded in relationships and associations and can be spread by word of mouth or over the Internet. With viral marketing, community citizens participating in energy efficiency programs voluntarily share their successes with their social networks.

Personal outreach. Residential programs also target builders, developers, and equipment contractors and distributors. To reach these audiences, program managers often hold industry-related trade-ally training events as a means to educate contractors about installation best practices, introduce new technologies, and create awareness of program offerings. While this channel is a premium from a cost perspective, the value is reaching an audience that directly influences consumer buying decisions. Contractors and trade allies are a crucial link to securing market adoption of new technologies.

A great example of this is the GasNetworks annual training event, sponsored by New England natural gas utilities. This all-day event attracts hundreds of HVAC and plumbing tradespeople each year and features the industry's best trainers to educate these contractors on how to properly sell and install high-efficiency space- and water-heating equipment. The event also includes a trade show for manufacturers and local distributors to introduce new high-efficiency equipment and promote their selling and energy savings benefits. Program administrators also use this opportunity to educate contractors about available consumer or contractor rebates offered on the same equipment being featured.

Participating in high-profile events that target residential consumers and/or contractors is another way to increase awareness and interest in energy efficiency. Home shows and community events are an excellent example of this opportunity. Program managers become very creative in promoting their message at these shows, which target both consumers and contractors.

Partnerships and delivery channels

Savvy program managers have learned that, to cost-effectively reach the large, diverse residential market, industry partnerships and established delivery channels are critical. Partnerships with other program administrators and Energy Star raise awareness of programs and provide a platform for common program content. Delivery channels move energy efficient measures into production in a consumer's home. Retailers and energy service companies play a key role in this space (table 5-3).

Table 5–3. Partnership and delivery channels

Channel	Partner or Delivery	Value
Energy Star	Partner	Consumer awareness Platform for programs
Program administrators	Partner	Creation of common programs Leverage program management dollars
Retailers	Delivery	Influence consumers at point of sale
Energy service companies	Delivery	Turnkey assessment Installation of measures
Trade allies	Delivery	Influence consumers at point of sale

Energy Star. The Energy Star program is a powerful partnership for program administrators. Energy Star offers program administrators a platform for consistent program messaging and developing localized energy efficiency initiatives for consumers.

The Energy Star brand itself is a great tool in the residential market. It provides consumers with a trusted label symbolizing energy efficiency. Consumers often base their trust and purchase decisions on the Energy Star label when buying new household products and appliances. In fact, 74% of households recognize the Energy Star label.[4]

Program administrators. Combining with other administrators in the same region to deliver joint programs allows program management dollars to be stretched further. When program administrators partner or collaborate, they work together to promote energy efficient technologies, create common energy efficiency programs, educate consumers, and promote contractor training and awareness. In this program delivery model, economies of scale are achieved. Organizations working together can reach their goals far more cost-effectively than any one organization can working alone. Program consistency is an important benefit from a regional effort as it reduces customer and contractor confusion and provides the ability to share program costs, such as marketing and administration.

Retailers. Because product accessibility is key for consumers, as rebates are very attractive to the residential market, working with retailers is imperative. A couple of traditional approaches can be used by program managers in working with retailers. One approach is to offer consumers rebates at the point of sale. In this model, program managers will work with retailers to train sales personnel on the availability of a rebate, the benefits of promoting higher-efficiency products, and consumer eligibility requirements. Depending on the rebate-processing design, this model may allow retailers to provide the consumer with instant or time-of-sale mail-in rebate forms.

Program managers may also work upstream to tap into the supply-side infrastructure of manufacturers, retailers, distributors, and others who have the opportunity to influence an end user's purchasing decisions. This is sometimes called a *push* strategy of the marketplace to get initiatives launched.

Energy service companies. Many program administrators secure energy service companies to deliver programs. Energy service companies have expertise in delivering energy efficiency programs to the consumer owing to their resources and experience in delivering programs at the implementation level. Program managers can tap into the knowledge and best practices inherent in energy service companies owing to their experience across many different programs across the nation.

Trade allies. Trade allies are a key source of information for consumers as they consider new construction or major equipment replacement. As such, developing partnership with contractor groups constitutes an important communication channel. Program managers find great investment value in educating contractors and installers not only on the various types and characteristics of efficient energy products but also on the proper installation and maintenance of these products. In turn, this prepares and supports trade allies when they advise customers who are investing in new-construction projects or retrofitting equipment.

Common Program Categories

Program managers combine efficiency opportunities with program types and tools to create a portfolio of residential program offerings. These offerings typically fall into the program categories of lost opportunity, retrofit, products and services, education and information, and research and development (table 5-4). For each of these categories, program managers will define the purpose and goals of the program, the target market and marketing approach, the target end uses, the recommended technology, the appropriate financial incentives (if any), and the delivery mechanism. Program managers will often further define these programs on the basis of low-income and non-low-income customers. Additionally, program managers will develop an implementation approach to maximize projected energy savings.

Table 5–4. Common programs and purpose

Program Category	Purpose
Product and service	Raise consumer awareness to support buying decisions
Lost opportunity	Capture benefits at time of construction or renovation
Retrofit	Encourage customers to proactively upgrade existing equipment with high-efficiency measures or equipment
Education and information	Educate on energy usage and increase energy efficiency awareness
Research and development	Explore new program concepts and technologies

Products and services

This category of programs includes lighting and appliance programs. Most lighting and appliance programs focus on Energy Star products and seek to raise consumer awareness of the benefits of purchasing energy efficient products.

With appliances, program administrators may also work to encourage higher efficiency standards for Energy Star–labeled appliances. These appliances include clothes washers, refrigerators, automatic dishwashers, consumer electronics, dehumidifiers, and air conditioners.

Program managers will want the design of lighting programs to support the development, introduction, sales, promotion, and use of energy efficient residential lighting products. The overall goals of the program may include increasing product availability, consumer acceptance, Energy Star product labeling, and adoption of new lighting technologies. Today's lighting program may target CFLs and fixtures, whereas tomorrow's programs may target LED lighting technology.

Lost opportunity

Lost opportunity includes the residential new construction and HVAC markets. Lost-opportunity programs capture energy efficiency opportunities at the time of construction, expansion, or renovation/remodeling or when customers make an initial purchase of equipment or replace failed equipment. If an energy efficiency measure is not implemented at that time, the energy savings opportunities are permanently lost, and the cost to retrofit later with higher efficiency is much greater.

A new-construction program can be designed to capture lost opportunities and encourage the construction of energy efficient homes by establishing the energy consumption and building performance guidelines used by the Energy Star qualified new-homes program. These programs are most effective if fuel blind, and they target customers and trade allies involved in the construction of single-family and multifamily dwellings.

An HVAC program should seek to raise consumer awareness of recent improvements in HVAC product technologies; to increase market share for Energy Star–labeled furnaces, central air conditioners, and air-source heat pumps; and to encourage customers to choose higher efficiency standards when purchasing HVAC equipment. This type of program may also include quality installation verification as part of the package. The market for this program is typically residential customers, along with HVAC contractors, technicians, suppliers, and distributors.

Retrofit

A retrofit program often includes a home energy audit, education, and assessments, along with incentives to encourage customers to install recommended measures, such as insulation, air sealing, and energy efficient lighting and appliances. The target market for this type of program is the residential customer or landlord in single-family or multifamily units.

Education and information

Programs in this category strive to educate residential customers on energy usage, to encourage positive attitude and behavioral changes with regard to usage patterns, and to increase customer awareness of energy efficiency programs. The audience may be residential customers, community groups, or school-age children. The goal of the program is to educate these target markets about the environment and energy savings benefits achieved through energy efficiency initiatives and programs. These programs may be delivered through educationally based program literature, online energy calculator tools, and community events.

Research and development

This category provides some flexibility for program managers to explore new concepts. A research and development program allows for evaluation and piloting of a new technology, market outreach measure, delivery mechanism, or incentive. For example, the research and development category would be a good place to test a new program based on LED lighting technologies.

Delivering energy efficiency solutions to the residential market is challenging given the size and diversity of the market. Program managers not only must develop a deep understanding of their market and the subtleties within the market but must also combine proven program design tools to deliver cost-effective energy efficiency programs.

Summary

Delivering energy efficiency to the residential market is challenging because of the size and diversity of the audience. Successful programs contain the design elements of program type, communication channel, and delivery channel and are grounded on a deep understanding of the market.

The direct install, audit and education, and rebate and loan programs are used by program managers individually or in combination to achieve their energy savings goals. Direct install programs comprise a core program type in which efficient technologies are installed in the customer's home. Residential consumers love energy audit and education programs, and these programs serve to raise awareness of the need for conservation. Program managers are challenged to provide this service in a cost-effective manner as direct savings are often difficult to derive from audits and education alone. Residential consumers respond to a rebate type of program more than to any other type. Program managers must carefully design the rebate amount to drive the appropriate behavior in a cost-effective manner.

Building on the program type is the communication channel. Residential program design must target the mass market. The traditional bill inserts and messages are a low-cost option that can be used to promote or augment a campaign. Program managers may decide to use higher-cost options, such as television, when there is a need to create broad awareness of a program. Fortunately for residential program managers, there are a variety of channels that can be used in combination to create the needed awareness and uptake on a program.

With the program type and communication channels in place, the final step is to determine the delivery mechanisms. Energy service companies offer turnkey assessment and installation. Retailers and trade allies provide a real-time connection with the buying consumer and can have a tremendous impact on the decision to purchase efficient technology. The delivery channels are strengthened by partnerships with other program administrators to create common programs offered in the region. In particular, Energy Star is a valued partner in the residential marketplace owing to the strong loyalty of consumers.

References

1 U.S. Department of Energy. 2004. *Buildings Energy Data Book*. Washington, DC: U.S. Department of Energy, table 1.2.3.

2 Policy Analysis Group, American Gas Association. 2008. *LDC Natural Gas Energy Efficiency Program Report 2007*. American Gas Association, Washington, DC, January 2008. prepared by Policy Analysis Group; published by AGA p. 13.

3 Chartwell Inc. 2008. *Utility Trends and Best Practices in Energy Efficiency*. Chartwell Inc., p. 11. published in USA.

4 Environmental Protection Agency, Office of Air and Radiation, Climate Protection Partnerships Division. 2008. *National Awareness of Energy Star® for 2007: Analysis of CEE Household Survey*. Washington, DC: U.S. Environmental Protection Agency, p. ES-1.

Chapter 6

Commercial and Industrial Energy Efficiency

The participation of the commercial and industrial (C&I) marketplace is critical to achieving energy efficiency goals. This sector represents approximately two-thirds of energy usage in the United States.[1] Therefore, it offers the greatest opportunity for savings.

Offering energy efficiency programs to the C&I market not only creates value for customers participating in the programs but also strengthens local businesses and contributes to a strong economy. Wise facility investments help to create a healthier environment and reduce overall costs. Furthermore, C&I consumers may find that incorporating the latest, most advanced energy efficiency measures into their facilities sets them apart from their competition.

Within the C&I market, the industrial and commercial sectors each represent nearly one-third of total energy consumption in the United States. Common industrial processes and systems, such as steam or process heating and motors, consume most of the energy and offer the greatest opportunity for savings.[2] This segment is not conducive to a mass-market approach because it will be unique by service territory and because of the use of application-specific industrial equipment and processes and a lack of industry-accepted test procedures.

The commercial marketplace features varied subsectors, including office space, food service, and hospitals. This market can be targeted with techniques like direct mail in the mass market, when focused on specific subsectors. The largest use of energy within commercial buildings is from lighting and space heating and cooling.

This chapter builds on the market barrier and assessment framework outlined in chapter 4, with a program design framework. Common programs for the C&I market are also reviewed. Program managers will want to combine the market assessment and program design framework to create programs best suited for their particular C&I audience and/or vertical market segments.

C&I Program Design

C&I program success is assessed on the basis of achievement of the targeted energy savings, customer satisfaction with the programs, and the cost-effectiveness of the program. Program managers are able to deliver successful programs building on a rich understanding of the market they are serving. This marketplace understanding, which is reviewed in detail in chapter 4, provides program managers with data on the various C&I segments, the size of these segments, and how these segments use energy. Program design encompasses the types of programs offered, how the programs are communicated, and how the programs are delivered (fig. 6–1).

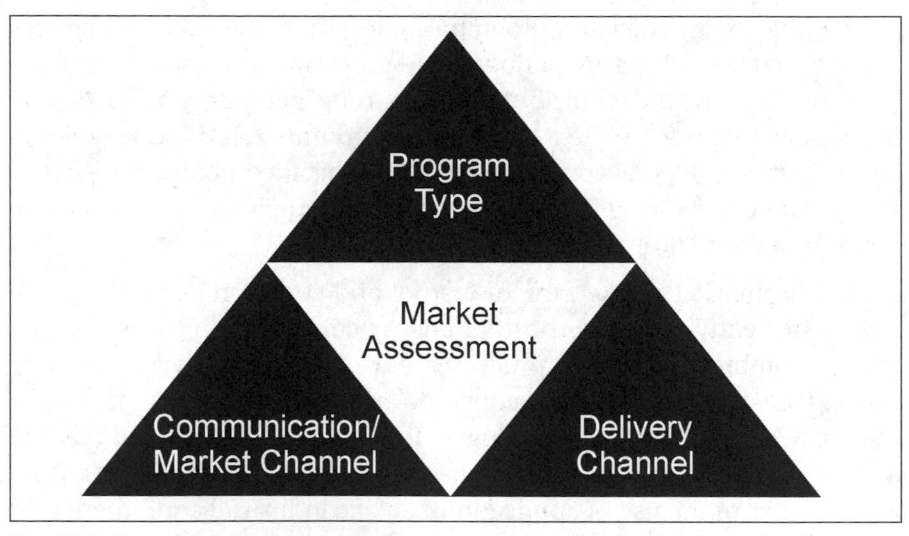

Fig. 6–1. Framework for program design

Program types

The types of programs for C&I energy efficiency fall into three categories: direct install programs; education and audit programs; and rebates and loan programs (table 6–1). Program managers will use these types of programs individually or in combination as they develop their specific portfolio of offerings.

Table 6–1. Overview of program types

Program Types	Definition	Audience
Direct install	Installation of efficient technologies with prescriptive or custom programs	Primarily retrofit; primarily small commercial; utilized for large C&I in targeted segments
Audits and education	Provide advice and education	End-use customers and trade allies
Incentives and loans	Financial incentive to influence energy efficient investment	Customers doing retrofit, new construction or equipment replacement; trade allies promoting products; upstream incentives to manufacturers

Direct install. Direct install programs are characterized by turnkey sales and the installation of efficient technologies targeting retrofit opportunities. The audience consists primarily of small commercial customers, but direct install may also be used for large C&I customers in targeted segments. Program managers may offer prescriptive or custom direct install programs.

Prescriptive programs feature prepackaged measures. Prescriptive programs offer a number of benefits, including targeting common end uses and offering predefined incentives. The application process for prescriptive programs is usually fast and easy for the customer. It also reduces costs for implementation because a single analysis has been done, resulting in an average energy reduction for the measure, as opposed to evaluating each application individually. For a measure to be made prescriptive, the volume should be large enough and the commonality of the resulting savings should be similar enough to have a manageable standard of deviation. Prepackaged measures for direct install include high-efficiency fluorescent lighting fixtures, lighting and controls improvements, motors and variable-speed drives, high-efficiency HVAC equipment, energy management systems, and refrigeration improvements.

For those projects that do not meet the criteria, there is still the opportunity to capture savings through a custom review of each measure. *Custom* refers to unique measures designed specifically for a particular customer. These can range from new technologies, to multiple-staged chiller plants, to customer industrial processes. For example, NSTAR worked with a technical school with old central plant equipment, consisting of chillers and boilers, that had surpassed its useful life. This formed the foundation for a series of improvements that included energy efficiency and mechanical infrastructure measures, resulting in an integrated, comprehensive, and efficient system.

Audit and education programs. Audit and education programs are popular with C&I customers and trade allies. Audits are intended to directly lead to energy conservation measures. Educational programs are behavior based and achieve energy savings by use of maintenance, control, and other operational strategies. Education programs may involve technical training (on technology or operation), program training, or even certifications.

Energy audits are used with C&I customers to identify and document potential energy conservation measures. Utilities still rely on on-site audits predominantly. However, online audits are becoming more prevalent, with 22% of utilities offering online audits in 2007.[3]

Energy audits can be targeted to specific programs. For example, the motor management pilot program by National Grid, an international engine delivery company, has a goal of promoting savings opportunities through improved motor management and a comprehensive approach to motor systems. In this pilot program, the customers' motor inventories were prescreened to identify potential candidates for the pilot program. A motor dealer or qualified technical representative performed an audit on the motor inventory of each candidate. The resulted in a customized report for the customer that detailed the new, replacement, and repair opportunities for motors, based on energy and dollar savings estimated, as well as on opportunities for variable frequency drives and controls.

Education is a valued offering in the C&I marketplace. Education may be delivered through classes offered by program managers or partnerships with various technical schools or regional organizations. Classes may cover a wide range of topics designed to encourage behavior such as ongoing facility benchmarking, facility commissioning, and retro-commissioning. Southern California Gas, for example, offers the Energy Resource Center, which offers an array of practical seminars, demonstrations, and consulting to help businesses find the most cost-effective and energy efficient solutions.[4]

Another tool being used to drive behavior through education is energy analysis software. Note that this is different from an energy management system or a building management system, which actively and automatically manages building systems. Energy analysis software instead provides customers with tools to allow them to understand where and when they are using energy so that they may better manage their usage and reduce costs. These tools are usually Web based and allow the customer to receive energy data, daily or in real time, in a user-friendly format. With more timely usage information, customers are better able to understand their load profiles, negotiate lower rates from energy suppliers, reduce their energy usage, and forecast energy costs. These tools can monitor energy from any source—including electricity, gas, oil, and water—and can be customized to track parameters such as temperature and humidity. The percentage of utilities offering this tool to C&I customers has grown significantly in recent years, from 17% in 2004 to 51% in 2007.[5] However, while popular with customers, energy analysis software is also difficult to tie neatly to reductions in energy usage. As such, program managers may offer this service in a fee-based manner.

Incentives and loan programs. Incentive and loan programs work to persuade consumers to invest in efficiency opportunities. Incentives are a powerful tool for a program manager and are widely used by program managers. Government-owned utilities are most likely to offer C&I customers incentives, at 71%, while co-ops are less likely, at only 27%. Most investor-owned utilities, at 58%, offer incentives. Heating and cooling related incentives are the most offered type, at 62%, followed closely by rebates for lighting, at 59%. Other incentive categories include motors, customer, industrial processes, variable-speed drives, and weatherization.[6]

Establishing the incentive amount is both an art and a science. First, program managers need a basis for determining the appropriate incentive levels. C&I program managers often look to the standards established by their partners at Energy Star and the Consortium for Energy Efficiency (CEE). Energy Star standards serve as a base level for energy efficient products. Higher standards apply for CEE's Tier 1 and Tier 2 measures.

Behind the scenes, CEE works with manufacturers, the U.S. Department of Energy, and program administrators across North America to review and define the Tier 1 and Tier 2 standards. Products in Tier 2 have higher efficiency than products in Tier 1. With the definition of efficiency standards in place, program managers then must decide incentive dollars appropriate to each level. Higher incentives will be offered for measures in the higher tiers of performance, for example Tier 2.

Rebates or incentives can be paid on a prescriptive basis, where the incentive is per unit. The measures associated with prescriptive incentives typically use established energy savings estimates that are based on values that have been studied over the years. Criteria for establishing incentives include unit cost, utility tariffs, project costs, and payback period.

A prescriptive approach and a well-designed application can reduce the amount of effort a customer needs to expend to apply for the programs. By identifying the measures most commonly used for lighting, cooling, and so forth, program managers reduce the amount of time necessary for customers to calculate the incentive. Program managers may even offer a spreadsheet online that allows the customer to input measure counts and other basic information to quickly calculate the incentive.

It is more challenging and cumbersome to define an incentive amount for custom measures. The process here involves gathering all the data necessary to evaluate an appropriate rebate amount. Data needed include the equipment specifications, costs, operating procedures, and hours of operation. Typically, custom applications are reviewed by internal technical experts that must fully understand and be comfortable with the engineering principles and assumptions to determine how the energy savings is to occur. This analysis leads to the determination of an appropriate rebate.

In addition to incentives, loans offer another means for customers to finance an energy efficient project. This type of financing product is less popular with the C&I market, with only 19% of utilities offering this tool. The interest rate offered on loans varies greatly, from zero to prime. Some loan programs are funded through the program administrator, while others are funded through banking institutions. The uptake on the financing program is very small as compared to rebate programs.[7] Regardless of overall volume, having a loan option may prove to be the factor that tips a number of projects forward because a borrower requires more attractive financing.

Communication channels

The commercial market consists of a variety of unique business types with unique value propositions. To reach customers in this market, program managers have a variety of communication channels available to them. Personal contact, through account managers, trade allies, or equipment vendors, often has the most impact in the C&I market. Mass-market communication channels, such as the bill insert and the bill message, have less of an impact in the C&I market owing to the unique needs of the various business segments.

This section provides a channel management assessment tool for matching communication channels to programs (table 6-2). The tool highlights the typical costs per transaction or per customer, again using the following categories:

- ***Low cost***—ranging from $0.00 to $3.00
- ***Moderate cost***—ranging from $3.00 to $10.00
- ***Premium cost***—greater than $10.00

Table 6–2. Communication channel assessment tool

Channel	Cost	Audience	Value
Direct mail	Low	Target audience	Focused program targeted to specific audience
Web	Low	Computer-savvy customers	Promote awareness; augment other channels
Print media	Moderate	Targeted audience	Promote awareness
Trade allies	Moderate	Customers evaluating energy options	Customers rely on trade allies to provide guidance
Direct sales	Premium	Trade allies; targeted customers	Promote awareness; educate on programs

Direct mail. Direct mail is a good communication channel for the C&I market because the promotional piece can be designed with a meaningful message about a specific program and mailed only to customers who would have an interest in those services. The message will often contain data, research, and case studies of other customers who have benefited from the targeted program.

Fourteen percent of NSTAR customers note that they heard about the company's programs through direct mail. This can be a cost-effective marketing approach when targeting a narrow audience.

World Wide Web. The Web is becoming a powerful communication channel for the C&I market as it is low cost and can support other channels. Program managers use the Web for education, augmentation of existing programs, promotion of programs, and lead generation.

Many program managers are enticing C&I customers to the Web by offering bill and energy analyzers, benchmarking services, energy-use profiles, and energy savings–tracking tools online to engage the customer. Program managers may custom design these online tools, or partner with firms, such as Aclara, that offer energy management applications. Aclara continues to see a growth of interest in Web-based management tools, as evidenced through its rapidly expanding major-utility client base. For example, according to Chartwell research, almost one out of four utilities now provides an online audit for C&I customers. Aclara's Richard Huntley, vice president of sales and development, has commented on the use of online management tools: "We expect this growth trend to continue as business customers, especially small and medium business, typically do not have in-house energy experts."[8]

Program managers may also provide energy educational material on the Web for C&I customers. Examples of this include a business library with searchable features, frequently asked questions, links to manufacturers, and other information. These Web sites can be augmented with links to vendors, announcements, ongoing energy publications, and even call centers.

Print media. Print media constitutes another tool for the C&I market. This includes program collateral and marketing in magazines and trade journals. Developing compelling collateral that can be provided to targeted audiences is necessary to effectively communicate programs. Advertising C&I programs through print media is challenging because of the limited channels.

To develop effective print media collateral that describes the programs, the program manager should consider designing from the customer's perspective. Print media can be targeted to various key C&I subsectors, such as property management companies, small businesses, or agriculture. Featuring brief success stories is one technique that creates more personalized material.

Print media also includes customer applications for entry into the program. Prescriptive programs rely on well-designed applications to gather relevant data. These data help in screening the applicant and provide the information needed for input into energy efficiency–tracking systems. Data that may be captured include a description of the equipment, including model number and location; a description of the business; the primary use; the end use targeted; and the specific program for which the customer is applying. Tables may be used as a part of the applications to make the form more user friendly; these tables can be pre-populated with equipment descriptions, so that the customer needs only to note the number of each type. Simply put, the application becomes an inventory checklist. The applications will secure data on the current equipment and the proposed changes so that estimated savings can be calculated.

Program collateral can also include promotional items, such as pens and corporate clothing. These can be targeted to a particular audience or idea. This form of collateral supports a broader, strategic message.

Advertising by use of paid print media can be effective, but it is challenging to determine the appropriate channels. Trade and business journals offer a channel to direct program advertising. The channels typically target a narrow audience, so the creative aspects must be geared to reach the audience with a meaningful message.

Trade allies. One of the most impactful communication channels for the C&I market is through trade allies. In fact, at NSTAR, 25 to 30% of C&I customers cited trade allies as the way they heard about the company's programs. Since this channel is valued by the C&I market, program managers target education and outreach to this group. This outreach may be in the form of education about installation best practices, introduction of new technologies, and creation of awareness of program offerings.

The vendor open houses hosted by NSTAR are a great example of trade ally outreach. These annual events feature presentations and updates on programs. This venue serves as an opportunity to enhance the relationship with key vendors while relaying important program information.

Sponsoring and participating in high-profile events that target the commercial market and associated trade allies is another way to increase awareness and interest in energy efficiency. Sustainable building conferences, trade association meetings, or building expositions are examples of events at which the message on the energy efficiency programs can be shared. Program managers become very creative in promoting their message at these shows, which target both consumers and contractors.

A key tool needed to effectively participate in expositions is a trade show booth. A booth is critical to both establish a corporate presence and communicate the proper message at the event.

Direct sales. Direct sales by account and program managers, marketing representatives, and energy engineers allows for direct contact with customers to identify their needs and cultivate the appropriate relationships. This is a vital communication channel in the C&I market. Armed with rich segment- or sector-specific information, these managers can increase penetration of the C&I sector or encourage repeat C&I customer investments in energy efficiency. Data from NSTAR indicate that 11% of customers receive communication on programs from account or program managers. Additionally, 17% of customers pursue further energy efficiency opportunities because of previous program participation. Account and program managers play a significant role in securing these repeat customers.

Importantly, to be most effective, account and program managers must be prepared with information on the business sector. This information may include data on the customers, issues, and key messages that can enhance the communication to this segment. Encouraging account and program managers to have face-to-face visits, supplemented by informational packets, case studies, and online Web tools, is a solid approach.

Partnerships and delivery channels

To cost-effectively reach the C&I market, effective partnerships and delivery channels must be created. For C&I, one of the most important channels is with program implementation contractors who not only assess leads but install measures. Other channels include trade allies, program administrators, and Energy Star (table 6–3).

Table 6–3. Partnership and delivery channels

Channel	Partner or Delivery	Value
Energy Star	Partner	Resources and tools for customers
Program administrators	Partner	Creation of common programs Leverage program management dollars
Program implementation contractors	Delivery	Turnkey assessment Installation of measures
Trade allies	Delivery	Influence customers at point of sale

With any of these channels, it is important to communicate the goals of the program and the type of measurement and evaluation process that will be used to determine the success of the program. Program managers can then use the feedback that they receive from partners and delivery channel personnel to improve the delivery process. Successful delivery of programs is a matter of adaptive management, meaning that the program can be adjusted on the fly. With rapid changes in technologies, tools, and customer expectations, programs must be adaptive.

Energy Star. The Energy Star program is a major partner to C&I program managers. From a C&I perspective, Energy Star provides utilities and other energy efficiency providers resources for the development of energy efficient and sustainable buildings, most notably the Energy Star building-rating system. Resources such as Portfolio Manager, an online comparison tool, can be utilized to bring buildings up to the Energy Star level. Additional resources include Building Design Guidance Checklist and Product Purchasing and Procurement, which aid in researching Energy Star–rated commercial equipment.

Program administrators. Partnering with other regional or national program administrators to deliver programs not only allows for common branding but also stretches marketing dollars and leverages combined budgets to influence manufacturing and other upstream market actors. When program administrators partner or collaborate, they work together to promote energy efficient technologies, create common energy efficiency programs, educate consumers, and offer contractor training and awareness, among other activities.

Program implementation contractors. Many program administrators secure program implementation contractors to deliver programs. The contractors come with a wealth of expertise in delivering energy efficiency to consumers. Using program implementation contractors provides program administrators with best-practice expertise, knowledge, tools, and proven delivery processes in delivering high-quality programs. Successful programs using program implementation contractors share certain common elements—most notably, clear partnership criteria, quality monitoring, and adaptive management.

Program managers and program implementation contractors perform best when the goals of the program are documented and there is a clear understanding on how performance will be evaluated. Mirroring these goals in the partnership or contract with the program implementer is important. Through alignment of goals and linkage between success of the program implementer and the same success criteria of the program manager, crossed goals are avoided, and a foundation for measuring performance is provided.

Quality delivery is expected but should be defined in the agreement. How will quality be measured and monitored? What is the escalation process when a customer complaint occurs? Defining the type and frequency of quality measurement and the feedback loops is important. For program managers to be successful, they need ongoing feedback on the program status and success.

Programs must adapt rapidly to adjust to changes in customer expectations, changes in technology, or other factors. Steve Cowell, chief executive officer of Conservation Services Group, a program implementation contractor, has suggested that "program managers and program implementation contractors must practice adaptive management in order to delivery the most successful programs."[9]

Trade allies. Trade allies are a key source of information and implementation of EE measures for C&I customers. As such, program managers are investing in outreach and education of contractors and installers not only on the types and characteristics of efficient energy products but also on the proper installation and maintenance of these products. In turn, this prepares and supports trade allies that advise customers on energy choices when customers are facing decisions—for example, on new construction and retrofit. This outreach also enables trade allies to properly install and maintain EE measures.

Program managers may use a series of tools to reach commercial trade allies. Some methods include vendor open houses and a vendor-only Web site. These venues offer trade allies updates on programs, policies, energy efficiency goals, and program status; announce upcoming events; and provide access to energy industry information.

Common Program Categories

Program administrators and managers build on their understanding of the market to design, develop, and deliver programs to C&I customers. Such programs target the unique needs of the customer segment that the program administrators are serving. The most successful programs are linked to a customer value proposition, use solid delivery channels, and have ongoing feedback mechanisms to allow for midcourse adjustments and quality assurance. Common program categories in the C&I sector include lost opportunity, retrofit, products and services, education and information, and research and development.

Table 6–4. Common programs and purpose

Program Category	Purpose
Product and service	Raise consumer awareness to support buying decisions
Lost opportunity	Capture benefits at time of construction or renovation
Retrofit	Encourage customers to install energy efficient measures
Education and information	Educate on energy usage
Research and development	Explore new concepts

Products and services

This broad category encompasses a variety of products and services that are self-supporting. Particularly in the area of C&I customers, it is helpful, from a marketing perspective, to wrap marketing messages and delivery around a specific service. Examples of services that might be offered are initiatives around motors, unitary HVAC, Energy Star benchmarking, or high-efficiency power supply.

The purpose of the programs delivered in the category of products and services is to increase customer awareness of and demand for high-efficiency equipment. Programs in this category may consist of a combination of marketing, education, and incentives to customers and trade allies.

In the Northeast, many program administrators offer regionally based programs vis-à-vis their affiliation with the Northeast Energy Efficiency Partnership. Regional initiatives leverage resources and coordinate program efforts toward common market transformation goals. An example of a regional program is the promotion of high-efficiency commercial unitary HVAC equipment and economizer controls that meet CEE's high-efficiency specifications. Regionally, program administrators worked to coordinate financial incentives, special promotions, and marketplace education, resulting in a wider range of products that meet CEE's Tier 2 specification and in increased sales of high-efficiency unites. In this case, the regional marketing outreach is promoted under the name Cool Choice.

Cool Choice has achieved a 12% market share for CEE Tier 2 equipment compared to all unitary HVAC equipment sales. This is an example of a program that will lead to market transformation, because the EPA and the U.S. Department of Energy adopted CEE's Tier 2 standard as the new federal standard beginning in 2010. Federal standards codify the efforts of program administrators.

Lost opportunity

The purpose of lost-opportunity programs is to educate and encourage customers to design features and select equipment today that will have an impact on energy consumption patterns for years to come. Therefore, it is important to encourage the optimization of the efficient use of electricity at the earliest decision stages in design and construction. The programs may combine rebates or incentives, technical services, and even commissioning services. These projects may involve new construction or a major renovation or remodeling of a building or may involve the purchase of primarily new equipment or the end-of-life replacement of fully depreciated equipment.

The lost-opportunity class of programs target all C&I businesses. These projects generally have longer lead times and involve a team of building developers, architects, and engineers. These projects are time dependent, and energy efficiency does not drive the schedule. The end uses targeted in lost-opportunity programs may include lighting, controls, motors, variable-speed drives, building-envelope measures, HVAC equipment, and compressed-air systems. These time-dependent projects involving the end-of-life replacement of depreciated equipment or the purchase of new equipment typically offer less comprehensive opportunities than do the overall building designs and tend to focus on a single system or piece of equipment. The customer in this case is the occupant or building manager.

The emerging trend in C&I energy efficiency programs is to focus on whole-building energy performance. With whole building performance, the objective is to maximize energy savings by aligning building operation, monitoring, and maintenance practices with design intent, and enable ongoing whole-building performance tracking. The potential savings estimates for the office sectors are between 10% and 20% through enhanced building operations and management practices.[11] Tools to support whole-building energy performance are available through Energy Star's benchmarking program and the nationally sponsored Commercial Building Initiative.

Retrofit

Programs in this category provide energy savings opportunities associated with existing electrical and mechanical systems in which the equipment being replaced continues to function but is outdated and energy inefficient. Program managers may offer financial incentives, along with technical services, in these programs. Financial incentives help with the high need in the commercial market for a short payback period. To raise awareness, program managers may offer contractors and equipment vendors training seminars. The end uses targeted in this type of program are similar: new construction, lighting, controls, variable-speed drives and motors, building envelope, HVAC, and compressed-air systems.

Program managers may further divide these programs into large and small businesses or other targeted subsectors. With small businesses, often program managers have contracts with program implementation contractors. These contractors may even have responsibility for marketing, outreach, and lead generation. Performance-based contracts may be used in retrofit programs.

Energy service companies are a critical trade ally as their core product is energy efficiency. They are in a unique position to leverage savings—from all fuels, water, maintenance, and so forth—to create a value proposition based on the customer's business and financial needs. Although a cash transaction is still the most common, most energy service companies also offer more creative financial vehicles, such as off-balance-sheet financing or performance contracting. Off-balance-sheet financing involves the leaseback of major equipment. Performance contracting involves financing and risk management, whereby the customer pays for the project out of a portion of the savings over time.

Education and information

Program managers will often offer programs under the category of education and information. The intent of such programs is to spur interest in the technologies and programs for which program administrators offer financial incentives or technical services. In this category, topics offered include facility benchmarking, energy management systems, commissioning, and retro-commissioning.

Technical training may be offered to support programs. The training targets installers, trade allies, and plant managers, among others. The training may be delivered by in-house personnel but is more often delivered through partnerships with other educational organizations.

Program managers may also do general outreach activities to promote new standards or tools. One example of such an outreach program is the promotion of Advanced Building Guidelines and Advanced Lighting Guidelines. These guidelines are intended to define the best practices in design, construction, and start-up of new and renovated C&I buildings. Promoting these building guidelines with various trade ally associations promotes awareness of these tools and increases adoption of the tools.

Program managers may also invest in Web-based educational tools. For C&I customers, these tools may offer information on energy efficiency technologies available in the sectors. Offering a customized portal for C&I customers that includes tools and promotion of various energy efficiency programs is a prudent investment in education and outreach for this segment.

Research and development

Research and development programs are used to pilot new program delivery processes or technologies. This is an important category to ensure that the pipeline of energy efficient technologies continues to be filled. Program managers always have an eye on programs needed for the future. With energy efficiency, development of energy efficiency technologies must continue to be encouraged. One way to do this is through the use of research and development programs, which serve to build the business case, for the value of a new technology or delivery process, not only for program administrators but also for other stakeholders, such as manufacturers or regulatory officials

Summary

Program managers can successfully design highly effective and popular programs by building on a market assessment. The program design that will include the program types, communication channels, and delivery channels will be configured to best serve the targeted audience.

The result is that offering C&I customers energy efficiency programs saves customers money and provides them with a more competitive position. Program administrators offering cost-effective, well-designed programs also gain from increased customer satisfaction. J.D. Power and Associates' 2008 business customer satisfaction survey has revealed that customers participating in energy efficiency programs had higher satisfaction than customers not participating.[12]

References

1 U.S. Department of Energy, Energy Information Administration. 2007. Electric Sales, Revenue, and Average Price 2006. http://www.eia.doe.gov/cneaf/electricity/esr/esr_sum.html

2 Consortium for Energy Efficiency. 2008. Work plans for 2008: Final 2008 industrial program planning document, p. 1.

3 Chartwell Inc. 2007. *Products, Services and Programs for C&I Customers Data Summary and Report 2007.* Chartwell Inc., p. 22

4 Southern California Gas Company. ERC at a Glance: Overview. http://www.socalgas.com/business/resourceCenter/ercOverview.shtml

5 Chartwell, 2007, p. 12.

6 Ibid.

7 Ibid.

8 Huntley, Richard. Personal communication with Penni McLean-Conner; August 11, 2008.

9 Cowell, Steve. Interviewed April 22, 2008.

10 NSTAR and E Source. 2008. Treating trade allies like family. Presented at the AGA/EEI Customer Service Conference, San Diego, CA.

11 CEE, 2008, p. 1.

12 J.D. Power and Associates. 2007. Small to medium size electric customer survey.

Chapter 7

Demand Response

The United States is experiencing new highs in peak demand. In fact, many regions reached demand levels that were not forecasted to occur for several more years. According to the North American Electric Reliability Council, peak demand will grow by 19% over the next decade; this demand is highly concentrated in the top 1% of hours during the year.[1] Demand response is a tool that helps shave the peak demand as an alternative to building power plants. As such, it is an important tool used by utilities and grid operators to effectively manage their grid under extreme stress.

Demand response is when energy users lower energy consumption during peak periods in return for receiving savings on their bills. Those savings can be a result of energy prices that are higher during peak hours or through payments made in return for specific actions such as reducing energy use to lower agreed-on usage threshold.[2] Research by the Federal Energy Regulatory Commission (FERC) indicates that about 37,500 megawatts (MW) of demand response potential exists across the United States. This represents the capability to reduce peak demand in most regions by between 3% and 7% of total.[3]

Utilities have a long tradition of designing grid systems that meet the expected energy demand from consumers. In fact, most utilities are generally required to build and maintain their systems to serve the highest expected total use of consumers, also known as the *peak demand*. However, weather, equipment malfunctions, and other unexpected events can create situations when the demand for energy exceeds the capacity to deliver; furthermore, new developments and redevelopment bring central air and high-tech loads to feeders not originally designed for this purpose. To avoid rolling blackouts, utilities and grid operators consider all available options to deal with capacity shortcomings—hence, the critical role of demand response. Bob Laurita, supervisor of demand resources at ISO New England, has observed that demand response is "a resource that can be rapidly deployed surgically in geographic areas where permitting would challenge traditional resources."[4] Demand response programs are successfully being used to manage the grid during peak times, and the success of demand response is leading to continued expansion of these programs.

Additionally, demand response is being viewed as a capacity alternative to meet future growth demands. The Brattle Group, a consulting firm in Massachusetts, has estimated that a 5% reduction in nationwide peak demand would eliminate the need for building and running 625 power plants, saving $3 billion annually.[5]

Finally, consumers are interested in participating in demand response programs, as a way to save money on their electric bills and to do their part in supporting the environment. Data indicate that the opportunity to save money is the overwhelming driver; over 90% of the time, it is the reason consumers choose to participate in demand response. Conversely, the reason consumers decline to participate is the concern that there is no guarantee of savings or their difficulty in understanding the program.[6]

Demand response is a proven and successful tool for utilities and for their customers. Bud Vos, vice president and chief technology officer of Comverge, has commented on the strengths of demand response:

> With rising energy use, the interest in demand response programs is unprecedented. Since 2003, COM verge's demand response resources, which started with 0 MW, jumped to more than 1,881 MW owned or managed in 2008, and the company continues to deliver more than 150 MW of equipment per year to utilities that are deploying their own demand response programs. It is a powerful tool to reduce load off the grid, improve reliability, and manage the environmental challenges we face today and in the future. Demand response is a cost-effective, rapidly deployed resource that provides benefits to utilities and customers—and ultimately the planet itself.[7]

Utilities are looking to both traditional, incentive-based demand response and time-based rate demand response to address their needs. This chapter provides an overview of both of these types of demand response and design elements.

Types of Demand Response

Currently, two broad types of demand response are offered to consumers: incentive-based and time-based rates. Both types drive customer response to prices or incentive payments. Both are designed to create changes in electricity use that are short term and centered on critical hours during a day or year when demand is high or when reserve margins are low.

Incentive-based programs

Incentive-based demand response programs are the time-honored method of load control and remain very popular. Incentive-based programs are offered in a variety of forms and are designed to respond to emergency or economic situations. Emergency demand response is where customers provide load reductions in response to generation shortfalls or transmission constraints. Economic demand response is where customers submit load reduction bids or simply respond to real-time prices. With each, to qualify for the incentive, customers must commit to reducing load within a specified period.

Commercial and industrial customers often participate in these programs as they may have load that can be curtailed. In fact, 80% of participants in price-responsive load programs have demands greater than 400 kW. This is in large part due to the minimum size or curtailment level required in order to participate under the programs.[8]

Residential programs are typically built around controls on air-conditioning units, electric water heaters, and pool pumps, which can be controlled with minimal impact on the consumer. Residential customers are aggregated by demand response providers who act as an interface with the end user and the utility or independent system operator (ISO) to deliver demand response capacity.

Although terminology and program names vary throughout the country, there are five main subtypes of incentive-based demand response programs (table 7–1). These programs provide incentives or direct payments to customers to induce curtailments when needed.

Table 7–1. Subtypes of incentive-based demand response

Program	Definition	Target Audience
Direct load control	Cycling of end-use devices, like air-conditioning or pool pumps	Residential and small commercial
Demand buy back	Customers choose to curtail upon request for event	C&I
Demand bidding	Customers bid load reduction into utility or market on advance basis	C&I
Interruptible rate	Rate for customers willing to curtail operations	C&I
Ancillary-services market	Ancillary services where customers receive payment for agreeing to reduce load when called	C&I

Direct load control. This is one of the most common subtypes of incentive-based demand response programs and can be achieved by cycling the appliances on and off or by using temperature offsets. Both of these options are targeted primarily to residential customers. Direct load control is mature, and data indicate that for each thousand customers, 1 MW of peak load reduction can be achieved.[9]

Cycling models are fundamentally based on controlling the air conditioner, to run half as much as it would normally run during the time period. Often the models use 50% cycling, which results in the customer's seeing an increase of two degrees Fahrenheit over two hours. Cycling models will also place controls on pool pumps where applicable. Duty cycling is flatter and more consistent and is a better application for longer control periods, of three to four hours.

Temperature offsets use smart thermostats, programmed with temperature settings. In this model, during a control event, a signal is sent to the smart thermostat, which then adjusts the temperature. This model achieves more immediate load reduction; however, it is best leveraged for short-term events, of less than two hours. The result of the immediate increase in the set temperature of the air conditioner is that the unit turns off; it will resume normal operations once the new set temperature is achieved.

Demand buyback. In this program, the customer chooses to respond to a utility's request for load curtailment for a specified period of time and price. Customers must have meters that measure usage during buyback periods. A base level of usage, from which reductions are credited, is defined and agreed on by the utility and the customer.

Demand bidding. This is a variation of demand buyback. In this program, customers are allowed to bid load reductions into the utility or ISO markets. If their bids are accepted, they are obligated to curtail. Customers can bid into this on a day-ahead market for hours prior to the day-ahead price release.

Interruptible rates. This rate-based program targets large industrial and commercial users, offering discounted rates or credits when they curtail their consumption as directed by their load-serving entity. These types of programs are attractive to customers who have either load that can be temporarily shut down, such as lighting, or distributed generation that can be powered up to provide continuous service during the interruption.

In these programs, notification is received from the load-serving entity. It is up to the customer to reduce load, usually for a financial incentive. This type of voluntary program pays credits to the customer for reducing load only when requested. Customer-driven and interruptible reduction programs represent the majority of commercial and industrial demand response offered by utilities.

Ancillary services market. Customers in this program are on contract to provide fast response to load reduction requests. This market is used to ensure grid reliability. Transactions in this market can take place a day ahead or an hour ahead of when customers actually use the electricity. Available for sale in this market are the following:

- Replacement reserves—generation that can begin contributing to the grid within an hour
- Spinning reserves—generation operating with additional capacity that can be dispatched within 10 minutes
- Nonspinning reserves—generation that is not operating but can be operating within 10 minutes
- Regulation service—generation that is operating and whose output can be increased or decreased instantly to keep energy supply and energy use in balance

Time-based programs

Time-based rates have been offered by utilities for decades as a pricing option to customers. Today, as a consequence of the adoption of the Energy Policy Act of 2005 (EPACT), there is increased interest in, discussion of, and design around new variations of time-based rates. In fact, research indicates that over 70% of utilities are planning or considering advanced metering infrastructure–enabled time-based pricing.[10] While time-of-use (TOU) rates are a mature price option, new time-based rates, such as critical peak pricing (CPP) and real-time pricing, are now being offered to consumers (table 7–2).

Table 7–2. Types of time-based rates

Time-Based Rate	Definition
Time-of-use pricing	Prices set for a specific time period on a forward basis
Critical peak pricing	Established time-of-use prices in effect except for critical peak load days, on which a critical peak price is in effect.
Real-time pricing	Prices vary based on the market

The adoption of EPACT was instrumental in advancing the discussion of time-based pricing programs. EPACT states,

> It is the policy of the United States that time-based pricing and other forms of demand response, whereby electricity customers are provided with electricity price signals and the ability to benefit by responding to them, shall be encouraged, the deployment of such technology, devices that enable electricity customer to participate in such pricing and demand response systems shall be facilitated, and unnecessary barriers to demand response participating in energy, capacity and ancillary service markets shall be eliminated. It is further the policy of the United States that the benefits of such demand response that accrue to those not deploying such technology and devices, but who are part of the same regional electricity entity, shall be recognized.[11]

EPACT further required utilities to offer time-based rate schedules and provide metering to customers requesting this by 2007. It required states to conduct investigations to determine when it would be appropriate to implement the new requirement. While EPACT is passive, it has spurred discussion between utilities and regulators on smart metering and demand response programs.

Most utilities have offered some form of time-based rates. Most common is a TOU rate targeted to large commercial customers. Today utilities are offering more forms of time-based rates, including TOU rates, CPP, and real-time pricing. While offering consumers more rate choices is a positive for utilities, there are technology and bill implications that must be evaluated and understood by utilities considering new time-based rate offerings.

TOU pricing. TOU pricing is well established, having been offered by utilities for more than 25 years. Large customers are typically the target audience for this offering, although some utilities do offer TOU pricing for small customers, such as residential customers. In this case, prices are set for a specific time period on a forward basis, usually not changing more than two times a year—for example, during season changes. The prices in this rate are designed to approximate real-time prices, based on the expected average costs of each part of the day and generally vary by season. Consumers are made aware in advance of the price, thereby allowing them to vary usage in response to the price.

From a customer perspective, TOU rates are more easily understood than other more complex rate forms, since the rates change for only two or three fixed daily time periods each season. Commercial and industrial customers are often the target audience for this offering because they generally have more load available for transfer from high-price periods to lower-price periods. The savings resulting from this transfer must be greater than the higher cost of the advanced meter for the customer to realize savings.

The technology associated with TOU pricing involves, at a minimum, a more sophisticated meter. This meter must be able to record reads in time increments and store them, as opposed to the traditional register read of a meter that progresses continually, from month to month.

CPP. CPP is a variation of TOU pricing wherein the established TOU prices are in effect except in certain peak load days or hours during which prices are greatly increased in relation to the price of generation. CPP combines the predictability of TOU rates with very limited use of a real-time market signal to customers when generation and delivery impose the greatest costs.

Gulf Power has been offering a CPP option to customers since 2000 with strongly positive results. Average demand reduction per hour during on-peak periods was 22% as compared to a control group. This rose to 41% during critical peak periods.[12] Other utilities are experiencing similar results in CPP pilot programs. For example, the myPower pricing pilot program by the Public Service Electric and Gas Company (PSE&G) tested customer response to pricing signals. This pilot program noted an on-peak period demand reduction of 47% during CPP events by participating customers.[13]

The technology associated with CPP requires not only a sophisticated meter, as with TOU, but also a customer notification system. CPP programs often offer customers technology to set their energy-consuming equipment automatically to levels that they have preprogrammed for each pricing period, to make it easy for them to participate and benefit.

Real-time pricing. Real-time, or dynamic, pricing is a relatively new offering that provides an option for the customer to pay whatever the real-time price is. This offering allows prices to be adjusted frequently to reflect real-time system conditions. Real-time pricing allows price changes as frequently as every hour. In this rate, price signals are provided to the user on an advance basis reflecting the cost of generation or electricity at the wholesale level.

There are some limited programs offering real-time pricing. Since 1992, Georgia Power has offered a real-time pricing program with over 1,600 customers, representing 5,000 MW. Georgia Power finds that the day-ahead program is most popular.

Customers benefiting from this type of pricing typically have sophisticated controls on their facilities, allowing them to manage usage on the basis of pricing signals. These systems typical provide real-time load profile data, real-time pricing information, graphical reports, tabular reports, data extraction, cost trending, load shaping, regression analysis, and automated threshold-based alarming, among other capabilities. Third parties are also offering load management services to the customer for a percentage of the savings. For many customers, this is an attractive option as they do not have the resources or expertise to comprehensively mine the opportunities associated with fully benefiting from real-time pricing. In this model, the energy management system is proprietary to the third party, who has the resources to monitor and manage the load in real time.

As with CPP, an advanced metering infrastructure is needed to support real-time pricing. This entails a combination of a TOU meter with a two-way, real-time communication system.

Demand Response Program Design

Demand response programs in some cases are very mature, having been in place for over 20 years. These mature load control programs, especially those controlling air-conditioning, have proven successful. All demand response programs share certain common design elements (table 7–3).

Table 7–3. Program design elements

Design Element	Consideration
Customer communications	Selection of communication channels for outreach and ongoing program feedback
Customer billing	Design of customer-billing processes for programs
Contracts	Contract design defining customer requirements and rules of program
Incentive and rate design	Design of incentives and rates to drive desired participation and meet business objectives for incentive-based and rate-based programs
Technical infrastructure	Design of infrastructure to communicate on events, measure and verify performance and support settlement
Measurement and evaluation	Design of measurement and verification to accurately account for program success

Companies intent on offering a demand response program need, as a foundation, a market assessment (as covered in chap. 4). As such, a first step comprises market research and load research to identify the target markets and the program design that will best meet that market. Since demand response is a more mature product offering, much experience and data have been garnered over the past 20 years from other implementations and can be used to develop this market and program design. In fact, many demand response vendors can provide a utility with detailed data on expected penetration rates and optimum target audience. This then just needs to be married to actual customer data owned by the utility.

Current demand response programs are categorized as voluntary and non-voluntary programs. These programs complement each other. Voluntary programs are a good way to introduce customers to demand response. With voluntary programs, customers receive incentives for participating but do not experience penalties for noncompliance. Starting customers in a voluntary program allows them to gain confidence in their ability to respond. Some of the customers may move to the more lucrative non-voluntary programs, which include penalties for noncompliance, having gained experience and confidence in their ability to respond consistently.

Customer communications

Customer notification is required for both incentive-based and price-based demand response programs. Additionally, customer communications occur through the bill.

The most successful designs for communicating about an upcoming peak time period or interruption will provide multiple channels for a customer to receive notification that range from telephone calls to e-mail or text messages. Some communications require a response from the customer, acknowledging receipt of the notification.

There are Internet tools that provide customers with real-time interval load data on their energy usage, allowing them to increase program participation and compliance because they are able to monitor and adjust their load reduction during an event. These Internet-based tools increase customer satisfaction because they enable customers to meet their load reduction goals and allow for analysis of schedule-driven energy efficiency measures, such as optimizing start/stop scheduling of chillers. These tools also allow customers to manage their energy usage in real-time and, at the same time, provide data to the system operator to verify demand reduction.

Utilities or energy service companies may offer Web-based energy data acquisition systems that allow the customer to receive energy data and load profiles in real time in a user-friendly format. Customers benefit from load profile data as this can help customers to negotiate lower rates, to more accurately budget and forecast energy costs, and to reduce energy usage. This can support increased customer satisfaction as customers can gain the financial benefits of incentives associated with demand response and have access to real-time data; consequently, customers can make operational changes and see the results immediately, as opposed to waiting until a bill arrives to see that something may not be operating correctly.

One tool to recruit customers to demand response is to offer customers advice on other energy management solutions, such as energy efficiency or renewables. Successful program administrators find that helping customers identify ways to respond to control events that fit their operations is a positive selling point. Program administrators may tap into the utility's strategic account managers to identify and consult with customers or offer audits.

Customer billing

The bill is an excellent communication channel and offers customers the linkage between the benefits of program participation and their energy costs. TOU is a mature rate, and utilities' billing systems are designed to support this rate. Real-time pricing and CPP present different challenges because of the reliance on real-time information. The impact that these rates have on the meter data management, as well as on the bill preparation and presentment, must be carefully assessed.

Real-time pricing and CPP present elements of increased amounts of data that must be processed and variability in the prices. Traditional utility customer information systems are not designed to support these types of rate structures. For example, in CPP, the peak price, in theory, can occur at different times of the day on different days. Logic may need to be added to accurately calculate the bill, along with associated logic to identify billing exceptions.

Utilities may append software to massage interval data-based billing computations and import the result into the final billing process. With interval data, the volume of readings or values increases exponentially. For example, if interval data are collected hourly for the month, there are 720 or more hourly values. This increases to more than 2,880 values if data are collected at 15-minute intervals. All of these data must be reviewed, validated, processed, and stored to produce an accurate bill.

Utilities are innovative in addressing these billing and data challenges. For example, Public Service Electric and Gas developed an adjunct system to perform the billing function for their pilot program.[14] By contrast, Pepco outsourced the billing for their time-based rate pilot program.[15]

Additionally, for success, adjustments in the printed bill are necessary to provide customers with meaningful feedback on their energy use by period of day. Bill presentment involves communicating to the customer about the multiple data intervals. This information may be communicated on the bill or in a complementary online presentment.

Contracts

A key element of demand response programs, particularly commercial and industrial programs, is the contract design. The contract defines the customer requirements and rules of the program. Some common elements of a contract include market rules, requirements of program administrators, and customer requirements. Definition of the amount of load controlled, rules associated with terminating the contract, duration before the customer is eligible to participate again, and restrictions on other rate incentives available while on this rate are specified, along with the costs associated with breach of contract.

Incentives and rate design

Designing the compensation system, or rate, is perhaps the most challenging and internally scrutinized element. For incentive-based programs, identifying the participant compensation that is in line with the value they provide is complicated. For time-based programs, designing the rate to drive appropriate behavior and reflect the market is a significant undertaking.

Incentive-based demand response targeting residential customers is becoming more prevalent. Estimates indicate that while residential customers use only 40% of the electricity, they could provide 53% of the demand response savings. Being better at managing their energy budgets, residential consumers have a higher price elasticity of demand, in economic terms.[16]

Utilities considering offering incentive-based programs must determine the appropriate level of incentives that are linked to the amount of load that the customer curtails and to market prices. The program may be designed to include penalties for event noncompliance. Program managers must design the appropriate incentives that attract the market and drive desired behavior, while meeting the business objectives.

Developing rates that provide price signals for TOU, real-time pricing, and CPP require analysis and decision. TOU pricing is a mature product wherein the rates are designed to approximate the market. The challenge today comes with rate design for real-time pricing and CPP.

The most accurate real-time pricing would be computed on a post basis when all the settlements have been finalized. However, such a tariff is unappealing to customers who have some degree of certainty about the price they face prior to the event. Hence, real-time pricing tariffs often use an advance price—day ahead, hour ahead, or near real time. For Georgia Power, which has a very mature real-time pricing program, the day-ahead form is much more popular where customers are able to actively schedule modifications when prices are expected to be high or low.[17]

When designing CPP tariffs, a determination of when to call a CPP event and the price of the CPP event must be made. The CPP value is not necessarily the market value of power. It may be a higher value that has been determined through participant price elasticity studies.[18]

Technical infrastructure

The technical infrastructure required to implement a demand response program varies depending on the type of program being implemented. For example, a mass-market direct load control program would require thousands of control devices and communication infrastructure capable of sending a local reduction signal to the devices. By contrast, an emergency demand response program for commercial and industrial customers might use telephones, pagers, e-mail, and/or fax notification and require that participants have interval meters so that load reduction could be verified.

- There are three main functions that must be satisfied by the technical infrastructure of a demand response program: ***Event communications.*** The technical requirements depend on the timing and content of the communication and could consist of a manual or automated system that might also directly control devices at the participant facilities.
- ***Measurement and evaluation.*** Infrastructure for measurement and evaluation usually consists of interval power metering capable of determining power usage for hourly or smaller intervals.
- ***Settlement.*** This function typically requires modification of existing settlement software to allow program participants to be paid, to receive a credit, or, in the case of a real-time hourly pricing program, to settle at the hourly price for their consumption.

Measurement and evaluation

Measurement and evaluation of program performance is vital for the program. Demand response resources are expected to perform and yield financial incentives associated with their reliable performance. The measurement and verification of demand response generally entails the use of interval electrical load data to establish the customer baseline usage during an event, which is subtracted from actual usage. The interval load data can be taken for the population or based on a statistical sampling of participants.

The strategy deployed depends on the type of program implemented. For the New York Power Authority Peak Load Management program, whole-facility metering is collected for all program participants. For mass-market programs, such as Xcel Energy's Save Switch program, end-use metering on a statistical sample is deployed. These projects can be conducted using a test-control experimental design or simply using average days or similar load days for the test group.

Evaluation should be done by an independent evaluator who does not receive compensation based on performance of the program participants. Increasing importance is placed on the demand response programs' performing as planned, and there are even markets, such as the ISO New England's forward-capacity market, that depend on this. Periodically evaluating these programs to ensure that they are designed and implemented as effectively as possible allows for optimization.

Summary

Demand response plays a key—and growing—role in the management of the grid for reliability, as an economic alternative resource to traditional generation and as a tool for consumers to save money. Data indicate that more and more utilities—well over 50% in 2006—are considering or planning demand response programs.[19] Traditional, incentive-based accounting still predominates, representing 60%–70% of demand response programs.[20] However, offering time-based tariffs is a growing trend.

When considering whether to offer demand response, it is important to start by using the market as a solid foundation. Completing a market assessment is vital to understand the size, potential, and interest various customer segments in demand response programs. Once an understanding of the market has been achieved, it is possible to design demand response program offerings, both incentive based and price based, to meet defined organizational objectives. In the design of programs, it is important to thoughtfully plan the technical infrastructure needed to support the demand response technology; the customer support systems needed for successful delivery, such as communication, outreach, and billing; and contracts, supported with measurement and verification to ensure accurate billing and settlement.

For utilities, the drivers behind demand response range from grid management, to customer satisfaction, to regulatory requirement. When asked about the drivers to offering customers with CPP and real-time pricing options Mike Sullivan, senior vice president of operations for Pepco, responded, "Customers today want help managing energy usage in order to save money and to reduce their carbon footprint. Today technology is available to offer this to customers and, on top, create a benefit to the utility from a grid management perspective. This is a powerful win-win combination that we could not afford not to pursue."[21]

References

1 Faruqui, Amad, Ryan Hledick, Sam Newell, and Johannes Pfeifenberger. 2007. *The Power of Five Percent: How Dynamic Pricing Can Save $35 Billion in Electricity Costs*. The Brattle Group, p. 1.San Francisco, CA.

2 McKinsey Consulting. 2002. *Demand Response and Advanced Metering Fact Sheet*. Washington, DC: Demand Response and Advanced Metering Coalition.

3 FERC. 2006. Staff assessment of demand response and advanced metering. Item No. A-5. July 20.

4 Laurita, Bob. Interviewed May 15, 2008.

5 Faruqui et al., 2007, p. 7.

6 Erikson, Jeff, Michael Ozog, Elaine Bryant, and Susan Ringof. 2007. Residential time-of-use with critical peak pricing pilot program: Comparing customer response between educate-only and technology-assisted pilot segments. Presented at the Energy Program Evaluation Conference, Chicago.

7 Vos, Bud, and Bill Mayer. Interviewed March 21, 2008.

8 Schwartz, Lisa. 2003. Demand response programs for Oregon Utilities. Report prepared for the Oregon Public Utility Commission, Salem, OR, p. 42.

9 Chartwell Inc. 2007. *AMI-Enabled Demand Response 2007*. Chartwell Inc.

10 Ibid., p. 8.

11 *Energy Policy Act of 2008*.

12 Schwartz, 2003, p. 39.

13 PSE&G's MyPower Pricing Pilot Program Wins National Recognition. Media Center Press Release; February 13, 2008; http://www.pseg.com/media_center/pressreleases/articles/2008/2008-02-13.jsp.

14 Chiu, Susanna, Pubic Service Electric and Gas, September 26, 2008. personal communication with Penni McLean-Conner.

15 Sullivan, Mike. Interviewed June 3, 2008.

16 McKinsey Consulting, 2002.

17 Borenstein, Severin, Michael Jaske, and Arthur Rosenfield. 2002. *Dynamic Pricing, Advanced Metering and Demand Response in Electricity Markets.* Berkeley, CA: University of California Energy Institute, p. 34. http://www.ef.org/

18 Ibid.

19 Ibid.

20 Goldman, Charles. 2006. *Energy Efficiency Funds and Demand Response Programs—National Overview.* San Francisco, CA: Lawrence Berkeley National Laboratory, Federal Utility Partnership Working Group.

21 Sullivan, 2008.

Chapter 8

Distributed Generation

At its simplest, distributed generation (DG) is the act of making electricity close to where it is used. The modern electric generation, transmission, and distribution grid was born in 1882, with construction of the first commercially viable power station, which was designed by Thomas Alva Edison. This was a DG installation that powered 1,200 streetlights all within a mere mile of Manhattan (fig. 8–1). The plant had the capacity of 600 kW powered by coal-burning steam boilers.

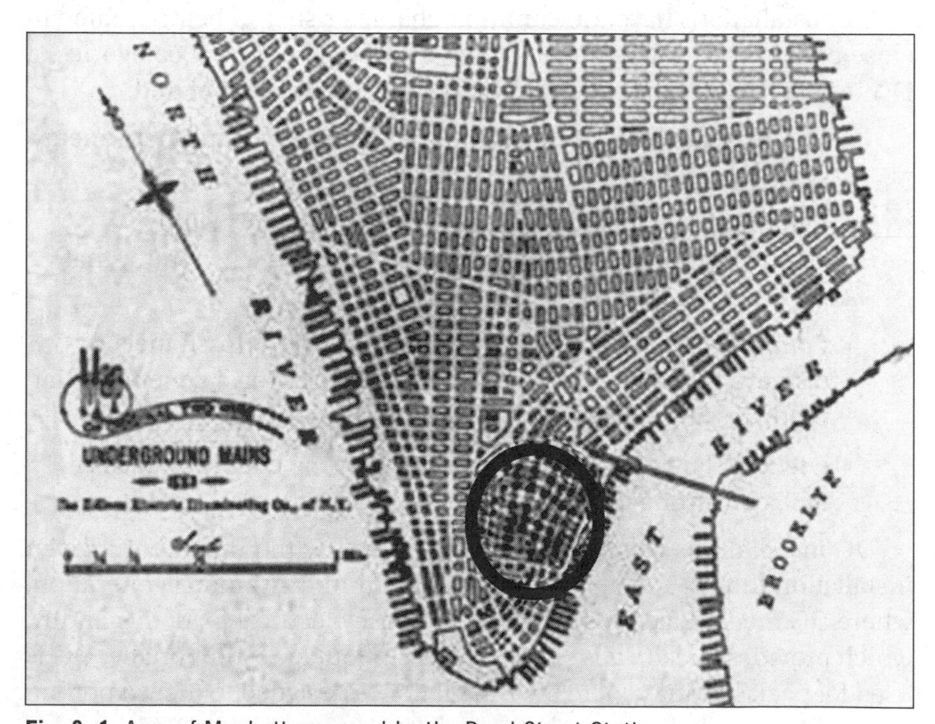

Fig. 8–1. Area of Manhattan served by the Pearl Street Station
(Source: IEEE Virtual Museum [http://www.ieee-virtual-museum.org/collection/event.php?taid=&id=3456876&lid=1])

Over time, innovations in the industry led to large-capacity generating stations and enabled the transmission of electric power over long distances. Life-cycle costs per kWh of electricity saw real decreases as the industry matured and demand for the power increased. Today's electric grid, characterized by power stations generating thousands of megawatts, transmission networks extending across the nation, and distribution networks connecting all households and businesses, is the result of innovation and optimization.

However, centralized power generation located far from load centers presents two significant challenges, each of which is the result of nature's thermodynamic principles. First, the most efficient generators are able to effectively convert about 40% of the thermal energy from combustion of fuel such as natural gas, coal, or oil to useful electric energy; the remaining 60% of the heat energy is typically vented to the atmosphere. The second challenge is that the act of transporting and managing the energy on the grid causes electric losses, predominantly heat loss through the wires and transformation losses associated with regulating voltage.

DG installations have the common characteristics of being located in close proximity to where the electricity is used. Four broad categories of DG installations exist, based on how the electricity is generated:

- ***Photovoltaic (PV).*** PV installations convert the sun's heat energy to electricity through a chemical reaction.
- ***Wind turbine.*** Wind turbine installations convert wind's kinetic energy into mechanical power, driving a generator that produces electricity.
- ***Combustion engine.*** Combustion engines burn fossil fuels, converting them into mechanical power, driving a generator that produces electricity.
- ***Fuel cell.*** Fuel cells chemically consume the energy of a fossil fuel, converting that energy into electricity.

DG installations can also be characterized by the need that a given installation fulfills. The applications range from continuous generation, where DG meets a facility's minimum or base electrical load, to standby, which provides backup generation in case of outage. Additionally, DG is used for peak shaving, where the installation is designed to meet a portion of a facility's peak demand, and for remote power, where DG is needed because there is not connectivity to the electrical grid.

DG can improve air quality and reduce greenhouse gas emissions if clean and efficient technologies are used. Many states are creating frameworks for expanding clean, efficient DG. The portfolio of clean and renewable DG includes reciprocating engines, microturbines, combustion gas turbines, fuel cells, PV, and wind turbines.

Interest and investment in DG is increasingly being driven by rising energy prices and interest in environmentally friendly energy options. This chapter provides an overview of DG technologies, a review of applications, and a summary of barriers facing DG.

DG Technology

DG "has the capacity to rewrite the economic relationship between traditional distribution utilities and their customers."[1] As such, utilities are gaining more expertise with DG technologies and more familiarity with their implications to utility grid systems.

A variety of DG technologies are commercially available today. These systems range in their efficiency and impact on the environment.

The most environmentally benign DG systems have zero emissions. The systems that fall into this category are fuel cells, PV, and wind turbines (table 8–1).

Table 8–1. Zero-emission DG systems

Type	State of Technology	Efficiency	Emissions	Application
Fuel cell	Emerging	25–50%	Zero	C&I
Photovoltaic	Maturing	5–15%	Zero	All
Wind	Mature	20–40%	Zero	All

Fuel cells

Fuel cell-powered generators represent another emerging technology. These systems are powered by an electrochemical energy conversion device. Fuel cell technology is similar to a battery in that an electrochemical reaction is used to create electric current. It produces electricity from various external quantities of fuel, on the anode side, and oxidant, on the cathode side. These react in the presence of an electrolyte. Generally, the reactants flow in and reaction products flow out while the electrolyte remains in the cell. Fuel cells can operate virtually continuously as long as the necessary flows are maintained.

There are four common fuel cell types: phosphoric acid fuel cells, molten carbonate fuel cells, solid oxide fuel cells, and proton-exchange membrane fuel cells. The fuel cells are named according to the type of electrolyte and materials used. The fuel cell electrolyte is sandwiched between a positive electrode and a negative electrode. Because individual fuel cells produce low voltages, fuel cells are stacked together to generate the desired output for the DG application.

These systems have efficiencies that range from 25%–60%, depending on the technology and the application. All fuel cell technologies have low emissions. Customers face some challenges with fuel cells, as the technology is still maturing. Maintenance of the fuel cell stack must be completed every few years, which can be costly.

There are several combinations of fuel and oxidant. A hydrogen cell uses hydrogen as fuel and oxygen as oxidant. Other fuels include hydrocarbons and alcohols. Other oxidants include air, chlorine, and chlorine dioxide.

Fuel cells are an expensive option currently at $4,000–$5,000 per kilowatt. The benefit is zero emissions.

PV

PV is a solar power technology that uses solar cells or solar PV arrays to convert sunlight directly into electricity. This technology involves assembling PV cells into flat-plate systems that can be mounted on rooftops or placed in other sunny areas. This technology is growing in popularity, in large part owing to financial incentives offered for the generation, including preferential feed-in tariffs, net metering, federal tax incentives, and state rebates.

PV panels convert light into electricity using the *photovoltaic effect,* which was discovered by Alexandre-Edmond Becquerel in 1839. Most applications of PV are interconnected to the utility grid; however, there is a smaller market for off-grid applications, such as remote dwellings, outdoor lighting for signage, and roadside emergency telephones.

PV systems are commercially available in sizes ranging from less than 1 kW to greater than 1 MW. The size of the system depends on the customer's available space for panels and budget. Sunlight is the fuel, so this technology has zero emissions, making it very environmentally friendly. The efficiency is 5%–15%.

Currently, PV technology is approximately two times the cost of grid power, but there are aggressive industry efforts to reduce manufacturing and installation costs. The U.S. Department of Energy's Solar American Initiative, for example, has several goals, one of which is to drive the cost of PV down to achieve grid parity by 2015.[2] Grid parity is the point at which the cost of PV electricity is equal to or less than grid power.

Over 90% of the world's PV installations are in Japan, Germany, and the United States.[3] Germany is the fastest-growing market owing to the attractive financial construct of the feed-in tariff.

Customers are attracted to solar because it is pollution free during use and because, with incentives, the payback is reasonable. The systems operate with little intervention after installation and have long life expectancy. Solar PV modules typically carry 25-year warranties, while inverters have 10-year warranties.

On the horizon is the further integration of PV into building design. An array may be incorporated into the roof or walls of a building. In addition, roof tiles with integrated PV cells are becoming available.

Wind turbines

Wind power is becoming increasingly popular because of its relatively benign environmental impact. The technology uses the wind for fuel. The wind turns the fan blades and powers the generator. Small-scale wind farms and individual units are often defined as DG.

Wind systems are available for residential applications or are grouped into small farms. The size range can vary from several kW to 5 MW. The efficiency is 20%–40%. There are no emissions associated with wind; however, siting wind projects can be extremely difficult because of concerns related to the visual impact of wind turbines.

Several DG systems are based on the use of combustion to generate electricity. The impact that these systems have on the environment depends on the fuel that is used for operation. Systems running on natural gas will have low emissions, but if the fuel is changed to fuel oil or diesel, there is a negative impact on the environment owing to high emissions.

Table 8–2. Combustion-based DG

Type	State of Technology	Efficiency	Emissions	Application
Combustion turbines	Mature	25–40%	Depends on fuel	All
Microturbines	Emerging	20–30%	Depends on fuel	Residential and small commercial
Reciprocating engine	Mature	25–45%	Depends on fuel	All
Stirling engine	Emerging	12–20%	Depends on fuel	Not commercially available

Combustion turbines

Combustion turbines are a very mature technology and can be fueled by a variety of sources, including natural gas, oil, and even dual-fuel options. The combustion turbine works by compressing incoming air to a high pressure. A combustor is used to burn the fuel, producing a high-pressure, high-velocity gas. A turbine then extracts the energy from the gas flowing from the combustion.

Consumers are generally business and industry and can choose from a variety of sizes ranging from 500 kW to 25 MW. The efficiency is 20%–45% and is dependent on the size of the system. The environmental impact can be very low when controls are used.

Microturbines

Microturbines are small combustion turbines derived from the turbocharger technology used in large trucks. These systems are being developed to also recover the heat from the exhaust gas to boost the temperature of combustion and increase efficiency. As an emerging technology, microturbines are relatively new to the commercial market. Their small size supports a market of small businesses and homeowners.

These systems can be fueled on natural gas, hydrogen, propane, and diesel and range in size from 25 kW to 500 kW. Their environmental impact can be very low but is fuel dependent. The efficiency is 20%–30%.

Reciprocating engines

This technology was one of the first DG technologies and is the most common, representing 75% of the world's generators.[4] The technology works by converting energy contained in a fuel into mechanical power.

The mechanical power is used to turn the shaft of an engine. A generator is attached to the internal combustion engine to convert the rotational motion into power.

This is a very mature technology and hence is the most affordable, reliable, and responsive. Reciprocating engines use commonly available fuels such as gasoline, natural gas, and diesel fuel but can also be powered by landfill gas or digest gas.

Customers can choose from a wide variety of sizes with this mature technology. The range is from a small residential backup generator, at 5 kW, to a large generator, at 7 MW. The efficiency depends on the application and ranges from 25% to 45%. The environmental impact depends on the fuel used, and emissions controls are required for NO_x and CO.

Stirling engines

Even though this technology was patented in 1816, it is an emerging technology with respect to DG. The Stirling engine is an external combustion engine that uses a sealed system with an inert working fluid, either helium or hydrogen. These systems are not commercially available. Stirling engines have the potential for very low emissions and have a low efficiency rating of 12%–20%. Natural gas is the primary fuel, but other fuel options are available.

Combined heat and power

Completing the DG technologies is combined heat and power (CHP). CHP installations are a composite category of DG. CHP installations typically use one form of raw fuel to create two or more useful forms of energy. A typical CHP system will have a prime mover, such as a combustion turbine, reciprocating energy, or a fuel cell, that produces electricity by powering a generator (table 8–3).

Table 8–3. CHP system

Type	State of Technology	Efficiency	Emissions	Application
CHP	Maturing	50–90%	Depends on fuel	All

Energy efficiency program administrators may want to consider expanding their programs to offer CHP to their customers as another way to achieve energy efficiency, since CHP can save energy by using what would otherwise be wasted by power plants to supply the customer's electricity. In fact, National Grid, a utility serving New England, offers incentives to residential consumers for the purchase of a residential CHP unit called *Freewatt*.

CHP represents a large potential contribution to the goals of most energy efficiency programs owing to energy savings. The scale of CHP's potential contribution is driven by its increase in the overall efficiency of energy use, since the use of waste heat at a customer site displaces both the reliance on electricity generated at lower efficiencies in the wholesale market and the use of boiler fuel for customers' heating requirements.

The CHP installation uses the hot exhaust gas from the prime mover to generate either steam or hot water. Steam is generated in a specialized boiler called a heat recovery steam generator (HRSG). High-temperature hot water (HTHW) is created using a specialized heat exchanger in which the hot gases pass over tubes, heating water under pressure, but not to such an extent where the water would be converted to steam. The steam or hot water in turn is used to heat or cool a building or is used to power process equipment, such as steam-driven pumps and other mechanical equipment.

When a CHP installation is contemplated, thermal energy, heating, cooling, domestic hot-water process, and electrical requirements are considered because the DG installation will be situated in close proximity to the load center. Figure 8-2 illustrates CHP's inherent ability to improve process.

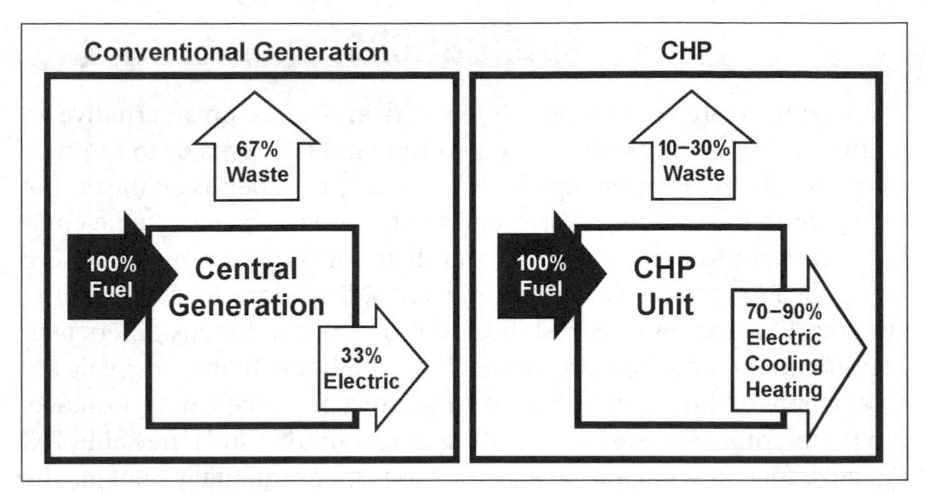

Fig. 8–2. Comparison of conventional generation versus CHP

Because, in a properly designed CHP system, more of the input fuel is converted to useful energy, less input fuel is used than in traditional, distant generation and localized combustion of fuels to produce heat and domestic hot water. Since less fuel is used, the utility expense is lower, and greenhouse gas emissions are also lower.

Biogen Idec in Cambridge, Massachusetts, is an example of a customer with an application for CHP. The challenge for this customer was reliability of electric and district steam supply to minimize the potential damage to research and production. Alternatives considered included investment in the distribution grids for electricity and steam, moving out of state, or installing a CHP system. The right answer for Biogen Idec was a CHP system using a 5.6 MW combustion turbine and HRSG, backed up by the existing electric and district steam grids. The customer cites the benefits of significantly reduced utility expense, reduced carbon footprint for the campus, and improved reliability.

Applications

Customers are increasingly interested in DG as an alternative to address rising energy costs and environmental concerns or to improve power reliability. The potential for the DG market depends on the cost of the systems, future energy costs, and the technology's energy efficiency. Customers wanting more control over their energy future and who have an application that is suited for efficient DG use are investing in DG systems. For example, interest in CHP is on the rise for customers with high thermal loads. Other customers need high reliability. Hospitals are an example of this, as they cannot afford to experience a power outage. Environmental concerns are creating a new market for renewable DG systems, such as wind and solar, which are environmentally benign. The market applications for DG are continuous generation, standby power, peak shaving, and remote power.

Table 8–4. Market applications for DG

Application	Definition	Business Drive
Continuous generation	Power running continuously, serving a base load	Economics
Stand by power	Power running as alternative	Reliability
Peak shaving	Power that meets a portion of facility load	Economics
Remote power	Power to meet total peak demand	Unavailability of grid

Continuous generation

These installations are designed to meet a facility's minimum or base electrical load, operating 8,760 hours per year. Continuous generation is when the DG application produces power on a nearly continuous basis. Customers considering this type of DG application would be comparing the cost of managing their own system to the competing grid prices. Customers must consider the maintenance and reliability of the DG system and the applicable tariffs with respect to standby rates.

The Medical Area Total Energy Plant in the Longwood Medical Area of Boston is an example of a customer with an application for continuous generation. The challenge for the customers in the Longwood Medical Area was that the existing electrical grid was strained to serve the current needs, and they foresaw significant, highly dense, highly energy-intensive growth in their fields of medicine and research. Alternatives considered for managing growth away from the current area included investing in additional capacity with the local utility and installing a large-scale CHP system. The right answer for the customers was continuous generation using reciprocating engines and HRSGs. This system provided the Longwood Medical Area with reliability of supply.

Standby power

Power outages are costly and present high risk for some customers. These installations are designed to provide backup generation in cases where outages are costly in financial or other terms. The most common standby power installation is an emergency generator sized to fulfill life and safety requirements. Customers who need to improve power quality or reliability are a market for standby power installations. In these situations, businesses install DG units to protect against the risk and cost of power outages. The customers in this market include grocery stores, hospitals, manufacturers, and banks.

The application of standby power meets customer's needs for high reliability. These systems can switch from grid power to standby automatically or with a delay. The sophistication of the standby power—that is, how quickly it takes over—is dictated by the customer's needs. Customers interested in this system understand and have monetized the financial impact of outages or have processes that are of a critical nature, such as hospitals and airports. The DG technologies used in these situations will depend on fuel access, time to start the generation, and permitting, among other considerations.

My father was an example of a customer who needed standby power to run a dairy farm. He did not need automatic switchover, as he had some degrees of freedom with respect to managing the herd. Nevertheless, he absolutely needed an alternative should grid power be unavailable, as the cows had to be milked, or else they would face serious medical issues. In his situation, he chose the most cost-effective system, powered by the farm tractor.

Peak shaving

These installations are designed to meet a portion of a facility's peak demands in order to minimize operating expense. Peak-shaving DG installations are common in areas where the local utility's rate structure has a ratchet demand component and the facility has a seasonally peaking demand profile. A singular demand will cause premium costs in all months.

Utilities have peak demand charges that reflect costs associated with building and maintaining the needed capacity to serve the peak usage times. There is a market for DG systems for customers who can greatly reduce or shave peak demand charges. In this case, the DG system will run intermittently, coinciding with peak demand times.

Remote power

These installations are similar to continuous generation in that the design parameters require continuous generation. However, the capacity of the installation is sized to meet the anticipated peak demand due to the unavailability of grid power.

Rural locations, which may face significant cost in connecting to the power grid, may choose to generate their own power on-site. This would include commercial establishments, such as ranches, or residential homes that are remotely located or not accessible to the grid, as on an island.

Barriers

DG represents a very small portion of total generation in the United States. There is increased interest in, research into, and application of the benefits of DG in supporting future energy needs in an environmentally friendly manner. Currently, there are barriers to rapid expansion of DG, including regulation, grid interconnection, and utility tariffs.

Regulation

The regulations that have an impact on DG range from siting and permitting, to environmental, to interconnection regulations.

Siting and permitting remain a key barrier to many forms of DG, especially to large systems, such as wind. While environmentally benign, wind is extremely challenging to site because of local concern about the visual impact.

Interconnectivity

It is critical that the safety and reliability of the distribution and transmission grid be maintained. As such, a variety of technical requirements are needed in order to appropriately interconnect. The utility and the customer will have to install a variety of relays and monitoring devices to ensure stability and protection of the grid and protection of the customer's site and generation. The cost and complexity of the interconnection depend on the size and scale of the DG system and the organization and local condition of the distribution system. The costs associated with interconnecting to and upgrading the distribution system can be high and are, in most cases, borne by the customer.

Interconnection to the utility's distribution system is a complicated and technical issue and must be looked at on a case-by-case basis, except for smaller applications. Utilities must assess DG interconnection requests from the perspective of system reliability and safety. Some states are providing interconnection standards that are helpful to consumers contemplating a DG system, but these standards are not yet uniform between states. Consumers considering a DG system will need engineering support to design the appropriate controls to meet required interconnection standards.

Tariffs

Tariffs are in place and new ones are being developed to address concerns of utilities, consumers, and other stakeholders. Tariffs provide a framework in which to expand DG in an effective manner. Tariff mechanisms in place today include the standby tariff, the feed-in tariff, and net metering.

The standby tariff is a mechanism that ensures appropriate utility compensation for building and maintaining grid connectivity for DG customers. DG customers interconnecting to the utility grid rely on the grid for power in case of a DG outage or during times of maintenance. The challenge for the utility is that distribution infrastructure must be designed to support the customer's total load, even though the load may be on the system for 10 minutes a year. The costs of providing backup power are often captured in utility standby tariffs.

The feed-in tariff provides customers with a guaranteed price on a multiyear contract. In this model, there is a separate meter that records the energy output to the grid. Germany introduced the feed-in tariff to support the expansion of PV. This tariff has been a powerful instrumental in Germany to create a new line of business of solar farming.

Another option is net metering, where the generation is first used to offset load within the facility and excess is then placed on the grid at a guaranteed price. Net metering uses one meter that goes forward and backward.

Summary

Delivering DG to consumers requires an understanding of the technologies, the applications, and the barriers. Interest in DG is increasing with the development of highly efficient, environmentally friendly technologies, rising energy costs, and the customer's needs for highly reliable power. A variety of technologies can be deployed, ranging from the very environmentally friendly (e.g., PV, wind, and fuel cells) to the more traditional (e.g., combustion turbine). Development and advancement in all of these technologies is rapidly occurring, offering customers increased efficiency and reduced emissions.

CHP is a composite category of DG that has the benefit of very high efficiency. This comes from its use of waste heat to generate steam or hot water. Energy efficiency program administrators may start incorporating this type of technology into program portfolios owing to the efficiency gains it promises.

Interest in DG will continue to increase. It is important for utilities and other energy stakeholders to build an understanding of the technologies and applications to ensure that these technologies are effectively integrated into the electrical grid. Program administrators must also consider whether and how to deliver DG solutions to customers.

References

1 California Energy Commission. 2002. Distrusted Generation Strategic Plan.

2 U.S. Department of Energy, Energy Efficiency and Renewable Energy. Solar America Initiative. http://www1.eere.energy.gov/solar/solar_america/goals_objectives.html

3 MarketBuzz2007: Annual World Solar Photovoltaic Industry Report. Solarbuzz reports world solar photovoltaic market growth of 19% in 2006. http://www.solarbuzz.com/marketbuzz2007-intro.htm

4 African American Environmentalist Association. Distributed Generation. http://www.aaenvironment.com/dg.htm

Part Three:

Optimize EE Performance

Chapter

Participate in Organizations that Advance EE

Program administrators are tasked with the achievement of specific plans, goals, and objectives. Program administrators also are responsible for effectively managing large energy efficiency budgets. The advantage to program administrators of working with other administrators, as facilitated by various regional and national organizations, is that the collective budgets and investment in energy efficiency can serve as a leverage tool. Active participation in key organizations can accelerate achievement of plans and goals.

Program administrators collectively represent a very large investment in energy efficiency. The combined U.S. and Canadian energy efficiency budgets total $3.7 billion.[1] In addition, as figure 9-1 demonstrates, these budgets are rapidly increasing as more and more states and provinces mandate funding for EE programs.[2] Budgets for both electric and gas efficiency are growing. Gas efficiency budgets increased 68% while electric budgets increased 14% from 2006 to 2007. Electric budgets far outstrip gas budgets to date, with $2.7 billion in electric budgets versus $416 million in gas in 2007.[3] Working together, program administrators are able to multiply the effect of their funding dollars and, at the same time, achieve greater energy efficiency for the public good.

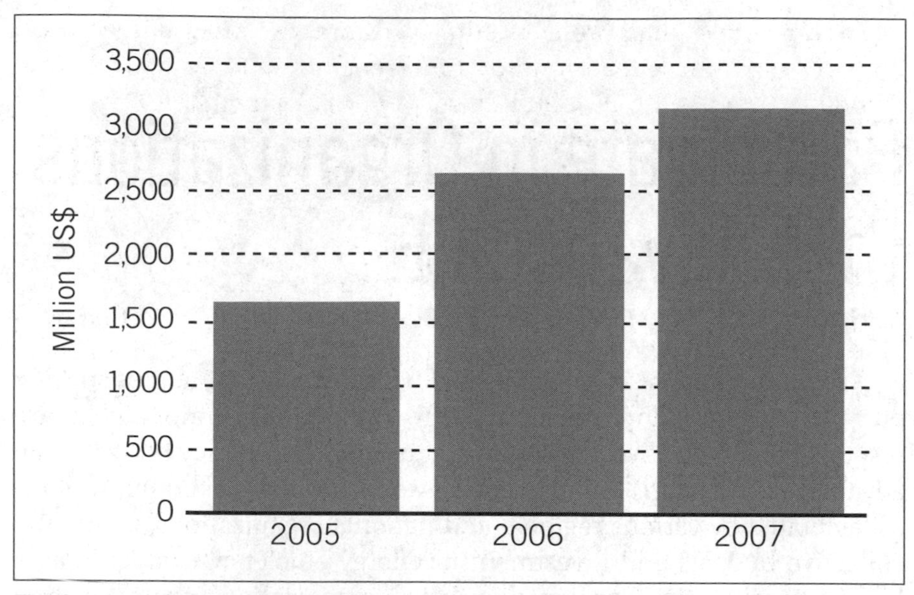

Fig. 9–1. U.S. efficiency budgets (Note: Data from 2005 do not include low-income and load management programs.)

Fortunately, there are many organizations that share a common interest in advancing energy efficiency. Understanding these organizations—who they are and what goals and objectives they pursue—is helpful to EE program administrators. These organizations can help program administrators achieve program goals and objectives. Program administrators must evaluate participation in these organizations in terms of their goals and the benefits of each specific organization.

There are many organizations in which to participate. Both time and budget preclude being productively involved in all of them. However, successful program administrators review the various organizations and match their time and resources to those that most strongly align with their goals and objectives.

More important, successful program administrators also fully participate in the organizations to which they belong. Carl Blumstein, director for the California Institute for Energy and Environment, serves on several boards for industry organizations including the Consortium for Energy Efficiency (CEE) and the American Council for an Energy-Efficient Economy (ACEEE). Blumstein emphasizes this high level of participation: "Gaining insight and best practices from other program administrators and from industry organizations is a contact sport. One cannot simply sit on the sidelines, pay the member fee, and expect to gain meaningful insight. Successful program administrators recognize the need to be on the playing field, fully engaged in the debate, conversation, and sharing of information."[4]

These organizations provide a variety of benefits, from advocating policy to completing research and development. All of them can advance specific programs, if goals are aligned and administrators extract the value through active participation.

This chapter reviews the benefits of participating in these organizations from a program administrators' perspective. The benefits of participation are discussed in terms of assessment criteria derived from best practices, including collaboration, leverage, and communication. This chapter also reviews the organizations that presently aid or advance EE.

Benefits of Participation

Organizations advancing energy efficiency provide a variety of benefits. Program administrators evaluating participation must look critically at their own program goals and objectives and identify gaps in achieving the objectives. Active participation in one or more of these organizations will aid successful program administrators in meeting targeted goals and objectives by bridging those gaps or enhancing their current activities and process. The benefits of participation include advocating policy, professional development, program advancement, research and development, and education. These organizations work on either a national or a regional geographic scale or both.

Advocating policy

Organizations operating at all levels—internationally, nationally, and regionally—may focus on advocating policy. Sue Coakley, executive director of Northeast Energy Efficiency Partnership (NEEP), has observed that "changing policy is the quickest and most effective way to advance energy efficiency success on a regional and national scale."[5]

Organizations advocating public policy will work to maintain a positive environment for energy efficiency policies by facilitating strategic alliances to develop and disseminate information. Policy advocacy involves outreach to inform policy makers about specific issues and opportunities to advance energy efficiency. Organizations that advocate policy complete research, collect data and perform analyses, and building and present the business case for adopting and implementing policies, such as appliance standards, to meet higher-level public policy goals. This information is used in position papers, in regulatory testimony, or to educate key policy makers and stakeholders. These organizations may also promote policies and programs by forming coalitions with other advocates sponsoring or presenting regional or national forums to discuss and highlight energy efficiency policy examples, accomplishments, and opportunities.

For example, NEEP completed a forward-looking impact study of energy efficiency in New England. The study showed that while New England has done much in the area of energy efficiency, there is much more to be done. In fact, the achievement of energy efficiency to negate load growth is possible. This study is often used and has been cited not only by NEEP but by others as policy is advanced to increase spending on energy efficiency. On a national basis, there have also been numerous independent studies to support the premise of using energy efficiency and demand response programs as a means to address load growth.

Success with policy advocacy requires creditability to have a recognized and respected voice for energy efficiency. Policy advocacy typically occurs with outreach to policy makers and influential stakeholders in preparation for and during regulatory or legislative proceedings or forums at the state or federal level to address generic policy issues.

Policy advocacy includes tracking of active and emerging policy changes and disseminating updates on current status. Supporting and working with organizations such as NEEP, that track and provide such information is a useful approach to building an at-a-glance understanding of issues impacting a particular service territory.

Policy outreach also entails education through conferences, Webcasts, and policy summits, to highlight emerging and current policy issues and to engage in discussion and debate of policy alternatives. These venues advance the thinking on and provide momentum to active and emerging policy issues.

Involvement in an organization advocating policy is important for program administrators but also presents internal challenges. It is important that interest be aligned between the program administrator and the organization. Wise program administrators establish that sponsorship and view participation not as a blanket endorsement of all organizational policy positions. This allows a program administrator to pick and support policies that align—while managing those that do not align—with internal company positions.

A further benefit to program administrators of being involved in an organization promoting public policies is that a more comprehensive understanding of alternative positions can be achieved. This can be helpful in fine-tuning internal positions. When a program administrator aligns, there is benefit externally from alignment with advocacy groups. Internal policy positions must be developed and aligned on key policy areas to advance policy, through an organization, that is in the best interest of the program administrator's organization.

Professional development

Organizations with a focus on professional development will typically provide an array of information and educational programs and outreach. This array will include conferences where professionals can gather, network, and share ideas and best practices. These organizations often publish journals and books. Many offer certification programs.

Professional development organizations are usually national in scope and serve members through local chapters. The local chapters develop ongoing program content that serves local members while supporting the overall organization.

Organizations focused on professional development also may provide scholarships to encourage new entrants into the field. With the increased demand for energy service professionals, the work of organizations focused on professional development is critical to future workforce and resource planning.

Staying on top of the rapid developments in technology and practice with respect to EE is important to energy service professionals. Several organizations are focused on providing professional development and knowledge transfer opportunities for practitioners in EE. Membership in professional development organizations can be attained individually or through corporate sponsorship.

Program administrators engage their staff in professional development. A corporate sponsorship may be established to leverage the educational opportunities for all of the staff or to underwrite membership for specific staff members. Organizations dedicated to professional development will often offer Webcasts, regional and national conferences, and publications.

Program advancement

Organizations that advance programs do so either through specific program development or in collaboration with other programs. Organizations may create unique programs, based on member input, that target member-specific gaps. The collaboration across programs is provided when an organization brings together the key partners from utilities, research and development organizations, manufacturing, and others in the public and private sectors. The organization provides a forum for members to discuss, network, and exchange information.

CEE is an example of an organization founded by utilities and efficiency groups to bring all efficiency programs together into alignment and leverage each other's efforts where consistent, national approaches are required. CEE has become the home of the efficiency program industry and is where program administrators come to network, exchange lessons learned, and work as an industry with other industries critical to the success of member programs.

CEE does an excellent job of creating collaboration across programs through its ability to bring together multiple stakeholders while focusing on making member efficiency programs more effective. An example of this is their Lighting for Tomorrow residential lighting-fixture design competition, a partnership on behalf of U.S. and Canadian efficiency programs, the American Lighting Association, and the U.S. Department of Energy. The market barriers to uptake of Energy Star–qualified fixtures were inadequate demand and inadequate supply. The partnership attacked both these barriers simultaneously: first, with a design charrette to introduce the efficient pin-based CFL technology to fixture designers (a charrette is a collaborative session in which a group of designers draft a solution to a design problem); and, subsequently, with annual design competitions that have evolved to focus on attractive, affordable interior and exterior fixture families and solid-state lighting. As a result, over 200 fixtures have been recognized by panels of independent judges and promoted by efficiency programs to local lighting showrooms and builders.

As program administrators broaden their reach for energy efficiency, more specific programs are needed to address more specific barriers. Several organizations have emerged that have a laser-sharp focus on specific industries and practices. These organizations bring together stakeholders who have a strategic interest in promoting a targeted industry or practice. For example, the New Building Institute focuses on improving energy performance in new commercial construction, with one of their more specific efforts being buildings under 100,000 square feet, in a program called *Advanced Buildings Core Performance*. Smaller-sized buildings are constructed more quickly, and energy savings opportunities are not large enough to permit detailed analysis; the Core Performance program was designed to provide a simple, stepwise format for architects and engineers to incorporate energy efficient design into this market sector. The associated prescriptive incentives facilitate decision making by the project team and reduce costs to both the project and program administrators.

Research and development

Organizations focused on research and development endeavor to fund and promote new energy technologies. These organizations leverage the expertise of scientists and engineers. Grants and membership are often the funding source for research and development efforts.

Organizations focused on research and development provide the value of sponsoring pilot programs and tests, thereby reducing the costs and risks associated with developing new technologies. The opportunity for members lies in the commercialization of technology and in the return on the investment in research and development.

Education

While not a singular focus of any organization, many of these organizations have an area of focus on education of key stakeholders, consumers, and policy makers. This education may be accomplished through initiatives, such as the Alliance to Save Energy's green schools and green campus programs, which "teach students from kindergarten through college ways to reduce energy use. Math and science based experiential curricula extend their reach beyond the classroom into the students homes and communities."[6]

The education may be classroom style for practitioners of energy efficiency, such as weatherization installers, builders, HVAC contractors, and facility managers. EE program managers will tell you that the purchase of high-efficiency equipment is just the first step. Subsequently, it must be installed properly and well maintained to achieve the maximum energy savings benefits. This often requires a planned educational outreach program to facility managers or plant managers, among others.

Education may come from forums such as conferences. Affordable Comfort Inc. (ACI) has a strong focus on education. Their mission is to "bring together the thinkers and doers who are committed to finding out how buildings really work. They solve technical problems, learn about new techniques, tools and materials, and build businesses."[7]

Program administrators will support organizations focused on education often to support service providers in the industry. To deliver energy efficiency, program administrators work with many stakeholders, including facility managers, plant managers, and HVAC installers. Advancing the education offered to industry-related professionals benefits the entire industry food chain. For example, if a plant manager is not aware of new technologies, who sells them, and why it is important to convince upper management to invest in efficient technologies, then the products never get purchased from the manufacturer. If an installer is not well versed in installing the latest equipment to meet new code requirements and/or achieve peak operational performance, then the end result is lost energy savings, poor performance, and skepticism—undermining enthusiasm for future technological purchases. If program administrators or account representatives lack the expertise and an understanding of the technologies, then their role as liaison to industry partners diminishes the ability of securing an emerging and self-sustaining marketplace. Essentially, everyone involved in the industry is a stakeholder, and education across all sectors is an essential component of achieving marketplace success.

Geographic scope

Some organizations provide a national scope. Such a wide scope allows program administrators to flex their leverage and market influence in the industry. Program administrators may choose to participate in nationally based organizations to advance specific policy areas or more effectively address building and appliance codes.

For example, an individual program administrator with ideas on improving Energy Star would have a difficult time securing a meeting with the U.S. Department of Energy or the EPA. More power, influence, and leverage are gained when several program administrators can provide common feedback and ideas. This warrants the prompt attention of groups such as the U.S. Department of Energy and the EPA.

Regional organizations play a powerful role in the advancement of energy efficiency. These organizations are focused on meeting the unique needs of a region and therefore can collaborate effectively and act quickly to address issues related to advancing programs. The work of a regional group is accomplished through committees convened by the organization to address common issues. Regional organizations may also play an advocacy role and are commonly viewed as yielding good value on the investment of a program administrator.

Review of Organizations

This section explores the organizations that are advancing energy efficiency today. Table 9-1 provides a quick reference summarizing the benefits of each organization.

Table 9–1. Organizational summary (N = national; R = regional)

Organization	Advocating Policy	Program Advancement	Professional Development	Education	Research and Development	Geographic Scope
Affordable Comfort, Inc. (ACI)				X		N
Alliance to Save Energy	X	X		X	X	N
American Council for an Energy Efficient Economy (ACEEE)	X	X		X		N
American Society of Heating, Refrigerating and Air Conditioning Engineers (ASHRAE)		X	X	X		N, R
Association of Energy Engineers (AEE)			X			N
Association of Energy Services Professionals (AESP)		X	X			N, R
Consortium fro Energy Efficiency (CEE)		X				N
Electric Power Research Institute (EPRI)					X	N
GasNetworks		X		X		R
International Facility Management Association (IFMA)				X		N
Midwest Energy Efficiency Alliance (MEEA)	X	X				R
New Buildings Institute (NBI)		X				N
Northeast Energy Efficiency Partnership (NEEP)	X	X				R
Northwest Energy Efficiency Alliance (NEEA)		X				R
Northwest Energy Coalition (NWEC)	X	X				R
Southeast Energy Efficiency Alliance (SEEA)	X	X				R
Southwest Energy Efficiency Project (SWEEP)	X	X				R
Trade associations	X	X			X	N

Affordable Comfort Inc.

ACI is a nonprofit organization established in 1986 with a mission to advance the performance of residential buildings through unbiased education. Their foundation is based on defining the best way to make homes energy efficient while maintaining health and safety standards for homeowners and apartment dwellers. This has evolved into the combination of home performance with building science. On this foundation, installation and construction best practices become the technical standards for green community projects.

This organization includes builders, HVAC contractors, home inspectors, designers, manufacturers, government, utilities, researchers, and educators. ACI is considered a foremost authority and is recognized nationally by home performance contractors for their signature training events.

Alliance to Save Energy

The Alliance to Save Energy was founded in 1977 as a nonprofit coalition of businesses, government, and environmental and consumer leaders. This organization was founded by Senators Charles H. Percy (R-IL) ad Hubert H. Humphrey (D-MN). The Alliance to Save Energy supports energy efficiency as a cost-effective energy resource under existing market conditions and advocates for energy efficiency policies that minimize cost to society and individual consumers. Their mission is to promote energy efficiency worldwide to achieve a healthier economy, a cleaner environment, and greater energy security. To achieve its mission, the Alliance to Save Energy undertakes research, education programs, and policy advocacy.[8]

The Alliance to Save Energy is successful because of its solid track record in unifying the public and private sectors to promote a sustainable energy future. The board of directors includes respected leaders from Congress and the chief executive officers of utilities and energy service companies.

American Council for an Energy-Efficient Economy

ACEEE, founded in 1980, is a nonprofit organization dedicated to advancing energy efficiency as a means of promoting both economic prosperity and environmental protection. ACEEE achieves its mission by conducting in-depth technical and policy assessments; advising policy makers and program managers; working collaboratively with businesses, public interest groups, and other organizations; organizing conferences and workshops; publishing books and reports; and educating consumers and businesses.[9]

American Society of Heating, Refrigerating and Air-Conditioning Engineers

The American Society of Heating, Refrigerating and Air-Conditioning Engineers (ASHRAE) was established in 1894 with a mission to "advance the arts and sciences of heating, ventilating, air conditioning and refrigerating to serve humanity and promote a sustainable world."[10] This mission is achieved through research, technology development, and education and training. This is a membership organization that engages energy efficiency industry partners and practitioners.

Association of Energy Engineers

The Association of Energy Engineers (AEE) is a nonprofit professional society established in 1997 whose mission is to "promote the scientific and educational interests of those engaged in the energy industry and to foster action for sustainable development."[11] AEE primarily works through local chapters, but also publishes a newsletter and journals. AEE offers scholarships to encourage new professionals to enter the energy management field.

Association of Energy Services Professionals

The Association of Energy Services Professionals (AESP) was founded in 1989 as a nonprofit association. It is a member-based association dedicated to improving the delivery and implementation of energy efficiency, energy management, and distributed renewable resources. AESP provides professional development programs and promotes the transfer of knowledge and experience.[12]

AESP provides national conferences and regional training opportunities. Additionally, AESP publishes papers on the latest energy services best practices and concepts. Its signature event is a national conference, which is well attended by program managers and administrators interested in learning from each other through networking, workshops, and general sessions.

Consortium for Energy Efficiency

CEE was founded in 1991 as a nonprofit public benefits corporation with a mission of working with members to advance high-efficiency products, technologies, and services. CEE members are ratepayer-funded energy efficiency administrators and their public stakeholders. CEE brings together energy efficiency program administrators from across the United States and Canada to discover through conversation credible, unbiased solutions to issues in the efficiency program industry. CEE is able to partner program administrators with other industries, trade associations, and government agencies and provides a forum to explore common interests, exchange information, and seek consensus with their colleagues, as well as with industry representatives.

For example, CEE works through 16 program committees of members. One of CEE's major accomplishments is to provide program templates, called CEE Initiatives, that allow efficiency programs across the United States and Canada to voluntarily align on specifications of high-efficiency products and services that their programs will promote locally. Together, efficiency programs are better able to influence the availability of high-efficiency products meeting the agreed-on CEE Tiers of high efficiency, making it easier to encourage customers to participate in their programs. National CEE initiatives address products ranging from residential (appliances, gas furnaces and water heating, lighting, and air-conditioning, including its quality installation) and consumer (electronics) to commercial (lighting, unitary air-conditioning, food-service equipment, building performance), and industrial (motor systems, processes, and water/wastewater facilities).

One of the side benefits of working together on consistency in promotion of high efficiency is the exchange of lessons learned and peer-to-peer exchange. Marc Hoffman, CEE's executive director, has noted, "Efficiency program administrators and managers marvel at the value of coming together to share experiences, design common strategies to overcome market barriers to their programs and work as an industry with the Energy Star program and supplier industries. When members pool their human capital, everyone can take away much more than they contributed."[13] This initiative defines a set of high-performance commercial kitchen equipment that members can deliver as a package to targeted food-service market sectors.

Electric Power Research Institute

The Electric Power Research Institute (EPRI) was established in 1973 as an independent nonprofit center for public-interest energy and environmental research. EPRI's mission is to conduct research on key issues facing the electric power industry on behalf of its members, energy stakeholders, and society. With EE facing the industry, EPRI is planning a reemergence to provide research and development on end-use efficiency technology. EPRI achieves its mission by bringing together members, participants, their scientists, and engineers, and other leading experts to work collaboratively on solutions to the challenges of electric power.[14]

GasNetworks

GasNetworks was established in 1997 as a collaborative of local natural gas companies serving customers throughout New England. Their mission is to work with governmental agencies and industry partners to promote energy efficient technologies, create common energy efficiency programs, educate consumers, and promote contractor training and awareness of ever-changing natural gas technologies.[15]

International Facility Management Association

The International Facility Management Association (IFMA), formed in 1980, is the world's largest and most widely recognized international association for professional facility managers. IFMA certifies facility managers, conducts research, provides educational programs, recognizes facility management degree and certificate programs, and holds the world's largest facility management conference and exposition.[16]

Midwest Energy Efficiency Alliance

The Midwest Energy Efficiency Alliance (MEEA) was formed in 2000 with a mission of advancing energy efficiency in the Midwest to support sustainable economic development and environmental preservation. MEEA's goals are to produce dramatic energy savings by inspiring active participation in MEEA's portfolio of programs and to deliver partner value, public benefit, and regional value by being the lead source on energy efficiency policy and programs. MEEA has positioned itself as a leader in raising and sustaining the level of energy efficiency in the Midwest by fostering market penetration of existing energy efficient technologies and promoting new technologies, products, and best practices, including renewable energy.[17]

New Buildings Institute

The New Buildings Institute (NBI) was established in 1997 as a nonprofit corporation. NBI works with national, regional, state, and utility groups to promote improved energy performance in commercial new construction. NBI manages projects involving building research, design guidelines, and code activities to ensure all elements of this chain are available for use by energy efficiency programs throughout the United States.[18]

Northeast Energy Efficiency Partnership

NEEP is a nonprofit founded in 1996. NEEP's mission is to promote the efficient use of energy efficiency in homes, buildings, and industry in the Northeast United States. This is achieved through regionally coordinated programs and policies that increase the use of energy efficient products, services, and practices. These programs and policies help to achieve a cleaner environment and a more reliable and affordable energy system. NEEP's core values are advocacy, collaboration, and expertise in implementation. NEEP facilitates regional initiatives of program administrators to increase marketplace availability and adoption of quality energy efficiency practices and technologies. NEEP's public policy outreach develops and maintains a positive political environment for energy efficiency and facilitates public policy projects and forums to advance specific policies.

Northwest Energy Efficiency Alliance

The Northwest Energy Efficiency Alliance (NEEA) is a nonprofit organization working to encourage the development and adoption of energy efficient products and services. Members of NEEA include electric utilities, public benefits administrations, state governments, public interest groups, and efficiency program industry representatives. NEEA works to create energy efficiency in the marketplace by helping program administrators to leverage their resources to achieve greater energy savings. NEEA also works to accelerate the emergence of energy efficient products and services and encourage their market adoption.

NEEA provides local utilities with marketing and training platforms that they can use to help their customers become more energy efficient.

NW Energy Coalition

The NW Energy Coalition is an alliance of more than 100 environmental organizations, civic groups, utilities, and businesses operating in the Northwest United States. Their mission is to advocate a clean and affordable energy future for the region by meeting all new demand with energy efficiency and new renewable resources, providing a full and fair accounting of the environmental effects of energy decisions, and protecting and restoring the fish and wildlife of the Columbia River Basin, as well as consumer and low-income protection and informed public involvement in building a clean and affordable energy future.[19]

Southeast Energy Efficiency Alliance

The Southeast Energy Efficiency Alliance (SEEA) was incorporated in 2006 as a subsidiary of the Alliance to Save Energy. SEEA promotes energy efficiency for a cleaner environment, a more prosperous economy, and a higher quality of life in the Southeastern United States. Its goals are to position energy efficiency as a viable tool for strengthening the regional economy and protecting the environment; to promote energy efficiency to increase electric reliability; to empower consumers at all income levels through education on the benefits of energy efficiency, including energy savings and quality of life; and to promote the development of a vibrant energy services industry throughout the Southeast and growing markets for energy efficient products.[20]

Southwest Energy Efficiency Project

Southwest Energy Efficiency Project (SWEEP) was formed in 2001 as an independent, nonprofit organization focused on advancing energy efficiency. It does this by collaboration with utilities, stage agencies, environmental groups, universities, and other energy efficiency specialists. SWEEP conducts studies and engages in policy advocacy. Although SWEEP's main focus is on electricity conservation, it also addresses other efficiency issues, such as fuel use.[21]

Trade associations

Trade associations are increasing their focus on energy efficiency. In the United States and Canada, the major trade associations are energy service companies and utilities, including the Edison Electric Institute, the American Gas Association, the Canadian Electricity Association, the Canadian Gas Association, and the National Association of Energy Service Companies. These associations collectively represent the interests of their members by advancing policy and providing services valued to members. These associations are all increasingly focusing their policy and services on EE. Program administrators may find it worth their time to participate in content or initiatives provided by trade associations on EE.

Summary

Program administrators keen on optimizing the performance of their EE portfolio will want to be involved in industry organizations. Clearly, there are many organizations in which to be involved. Program administrators and managers must assess these organizations to determine which provide the best value for the investment. The investment will be in the form of annual dues and active participation through committees, workshops, and boards.

In assessing the relative value of various organizations, program administrators may want to consider the viability of the organization. Organizations performing the best exhibit clarity of mission. Such an organization has a clear mission and clear strategies to achieve it. In addition, the organization has a strong board of directors that demonstrates clear support for the organization; the board of directors is diverse, understands the organization's mission, and is committed to taking the organization to the next level. The organization demonstrates and understands its own value in collaboration with other stakeholders, including both public- and private-sector companies and organizations. The organization recognizes the leverage made in advancing EE and wisely applies that leverage to move ideas and initiatives forward. Many of these organizations have successfully brought new technologies or practices to the marketplace by leveraging their member's resources and credibility. Furthermore, the organizations provide valuable communications, playing a vital role in researching technologies and practices and disseminating information that is helpful to program administrators.

Program administrators are wise to actively participate in both regional and national organizations. Deciding which organizations to participate in is a key decision for program administrators. Once the decision has been made to participate in an organization, program administrators need to ensure that their representatives are fully engaged. Gaining insights to optimize EE programs is definitely a contact sport, requiring engagement and active participation. Active and successful participation requires both time and budget resources.

References

1 Consortium for Energy Efficiency. 2008. CEE 2007 report: Energy efficiency programs, p. 7.

2 Ibid., pp. 8–9.

3 Ibid., p. 7.

4 Blumstein, Carl. Comments from CEE board meeting, June 2008.

5 Coakley, Sue. Interview, February 7, 2008.

6 Alliance to Save Energy: Promoting Energy Efficiency Worldwide. http://www.ase.org/

7 ACI: Advancing Home Performance. http://www.affordablecomfort.org/

8 Alliance to Save Energy: Promoting Energy Efficiency Worldwide. http://www.ase.org/

9 American Council for an Energy-Efficient Economy. http://www.aceee.org/

10 ASHRAE. http://www.ashrae.org/

11 Association of Energy Engineers.http://www.aeecenter.org/about/body.cfm

12 Association of Energy Services Professionals. http://www.aesp.org/

13 Hoffman, Marc.Interviewed by Penni McLean-Conner on February 5, 2008.

14 Electric Power Research Institute. http://www.epri.com/

15 GasNetworks. http://www.gasnetworks.com/

16 International Facility Management Association. http://www.ifma.org/

17 Midwest Energy Efficiency Alliance. http://www.mwalliance.org/

18 New Buildings Institute. http://www.newbuildings.org/

19 NW Energy Coalition. http://www.nwenergy.org/

20 SEEA: Southeast Energy Efficiency Alliance. http://www.seealliance.org/

21 SWEEP: Southwest Energy Efficiency Project. http://www.swenergy.org/

Chapter

Evaluating Programs

Increasingly, EE and DSM programs are being considered an energy resource, just like generation, and consequently are being built into integrated resource plans to meet forecasted load. The challenge, though, with EE programs is that most are unmetered resources. The impact that the programs have in aggregate on meeting energy demand must be reliably ascertained through evaluation. Hence, great emphasis is placed on evaluation to validate program performance.

Evaluation is the systematic testing, in the field, of the assumptions used in planning. Evaluation feeds the EE life-cycle process from the perspectives of resource planning, program objectives, and program design (fig.10–1).

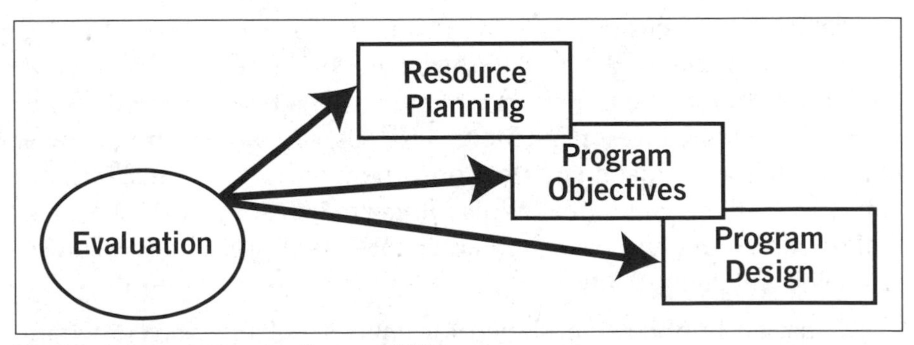

Fig. 10–1. Impact of evaluation on EE life-cycle process

Evaluation ultimately achieves two goals: improvement and accountability. From an improvement perspective, evaluations assess program performance and implementation effectiveness. The lessons learned from the evaluations are incorporated into new programs and are used to determine program objectives. From an accountability perspective, program administrators rely on evaluation to validate and document program results and ensure that programs are achieving program objectives and meeting resource-planning requirements.

Rigorous evaluation ensures that programs are cost-effective and that savings are sustained over time. Evaluation assesses whether a program achieved its goals, such as being a reliable energy resource. Of particular importance, program administrators rely on evaluation to demonstrate fiscal responsibility in the expenditure of public funds on EE programs.

The National Action Plan for Energy Efficiency has identified four types of evaluations: impact, process, market effects, and cost-effectiveness.[1] Impact evaluation quantifies the direct and indirect benefits of the program; this includes the energy savings, the demand savings, the emissions reduction, and other nonenergy benefits, such as water savings or improvement in quality of life. Process evaluations are used to assess the program performance; these evaluations are used by program administrators to identify improvement opportunities in program design and implementation. Market-effects evaluation assesses how the overall supply chain and markets have been impacted by the program; for example, it is important to evaluate whether a program or end-use technology has achieved market transformation, at which point the technology has become commonplace and is used by consumers without the need for any additional incentives. Cost-effectiveness evaluation quantifies the costs of the program and compares the costs to the benefits of the program; several types of cost-effectiveness tests are used, often at the direction of regulatory bodies.

Evaluations are costly, generally constituting between 1% and 15% of an annual program budget.[2] With the advent of EE being considered as a supply resource, the importance of evaluations has increased, as has the corresponding investment. Monica Nevius, senior program manager of research and evaluation for the Consortium for Energy Efficiency, has commented that "program administrators are investing time, attention, and resources to refine evaluation procedures to improve their accuracy, reliability and defensibility."[3]

Evaluation protocols developed for energy efficiency are providing the framework for evaluating demand response and renewable-energy programs, although there are some differences. For example, the time period used to evaluate energy efficiency is typically longer than the time period to evaluate demand response.[4] Efficiency and evaluation professionals have "addressed and provided guidance on key evaluation issues that need to be addressed in the evaluation of demand response, renewable energy and climate change programs."[5]

This chapter begins with an overview of the evaluation process, followed by a review of the types of evaluation. The primary focus, though, is on impact evaluation, by which the various impacts of EE programs, such as energy savings and avoided emissions, are determined. The other evaluation types—process, market effects, and cost-benefit analysis—are also reviewed. Evaluation is a mature process, and as such, a wealth of information is available and was used to develop this chapter. Key documents that were especially useful include the following:

- *Model Energy Efficiency Program Impact Evaluation Guide.* National Action Plan for Energy Efficiency.
- *EERE Guide for Managing General Program Evaluation Studies,* prepared for the U.S. Department of Energy, Office of Energy Efficiency and Renewable Energy
- "The Integration of Energy Efficiency, Renewable Energy, Demand Response and Climate Change: Challenges and Opportunities for Evaluators and Planners," by Edward Vine
- *How Do We Measure Market Effects? Counting the Ways, and Why It Matters,* by Elizabeth Titus, Monica Nevius, and Julie Michals

Evaluation Process

At the highest level, the evaluation process consists of four steps (fig. 10–2).[6] The process starts with the definition of the scope. The design of the evaluation supports the scope. Management of the evaluation ensures that all parameters defined in the scope are delivered. The final aspect comprises dissemination of the results and application of the lessons learned.

Fig. 10–2. The evaluation process

Define the scope

Successful program administrators recognize the importance of taking the time to carefully define the scope of the evaluation. The scope drives all other steps of the process. A scope starts with a clear purpose statement. With a well-articulated purpose statement, one can define the objectives, identify the resources, establish a time frame, and develop a budget for the evaluation.

It is helpful, in defining the scope, to start with the end result in mind. This means identifying the program decisions that need to be made as a result of the evaluation. For example, with a program featuring a more mature end-use technology, program administrators may consider whether to continue, adjust, scale back, or discontinue the program entirely. Rebate program administrators may want to ascertain whether the rebate is appropriate as is or, otherwise, whether it should be increased or decreased.

With a clear vision of what success in the evaluation looks like, specific evaluation objectives can be defined. The objectives can range from confirming the energy savings associated with the program to identifying program barriers or measuring customer satisfaction. As an example, a recent statewide evaluation completed in Massachusetts on the pilot program using the behavioral tool PowerCost Monitor defined the following primary objectives:

- Customer response to different PowerCost Monitor price points
- Customer perceptions of the value of PowerCost Monitor relative to other energy efficiency services
- Short and long-term behavior changes among participating customers
- Energy savings attributable to the pilot programs
- Success of different power-cost marketing strategies

Identifying and earmarking the resources needed to complete the evaluation is part of the scoping phase. These resources may include administrative staff associated with managing the evaluation and budgeted dollars for the study. From an administrative perspective, staff time will be needed to plan the evaluation, secure an evaluator, evaluate the results, and disseminate the results of the study. The budget associated with the evaluation will be predicated on the type of evaluation, the objectives, and the level of defensibility required.

Defensibility is the ability of an evaluation to stand up to scientific and regulatory criticism. Defensibility is an important construct as it has a heavy impact on the cost of the evaluation. The level of defensibility is assessed on the basis of the validity, reliably, accuracy, and bias associated with the evaluation. Validity is a judgment of the scientific standards that have been applied to the research design to give it credibility. Reliability describes the degree to which you would get the same results if you were to repeat the evaluation with the same design. Accuracy reflects the correspondence between the measurement made on an indicator and the true value of the indicator. Bias is the extent to which a measurement or sampling method underestimates or overestimates a value. Combined, these attributes determine the strength of defensibility. Program administrators, for example, tasked with defending the program to recover program costs, will want a high level of defensibility of the evaluations.

When defining the time line, evaluators must consider when decisions need to be made with respect to a program. This often defines the end date of an evaluation. Even though the end date is often not negotiable, owing to programmatic or regulatory constraints, it is important to ensure that adequate time is built into the project time line for the administrative tasks necessary upfront to hire an evaluation expert and for the quality-assurance review of the results when the evaluation has been completed.

Design the evaluation

With the scope defined, the next step is to design the evaluation. The elements of the design will be important components in the request for pricing (RFP) and in the selection of an independent evaluator. The components include selecting a type of evaluation and the program logic model. In addition to the model, the design will define the types of data to be collected, the information needed for the evaluation report, and the quality-assurance review process.

The types of evaluation are impact, process, market effects, and cost-effectiveness, each of which has a unique purpose. Sometimes these evaluation types are used in combination (table 10–1).

Table 10–1. Types of evaluations
(Source: Schiller, 2007, Model Energy Efficiency Program Impact Evaluation Guide [http://www.epa.gov/cleanenergy/energy-programs/napee/index.html])

Type	Phase at Which Implemented	Description
Impact	Implementation and/or post-implementation	Determines impacts and co-benefits that directly result from a program
Process	Implementation	Assesses how efficiently a program was or is being implemented
Market-effects	Implementation and/or post-implementation	Estimates a program's influence on encouraging future DSM projects because of changes in the marketplace
Cost-effectiveness	Implementation and/or post-implementation	Quantifies the cost of program implementation and compares with program benefits

The logic model diagrams the sequence of events that are intended to produce the results sought by the program. A well-designed logic model forces one to consider what the program is trying to accomplish and how well the activities produce the desired output, which in turn produces program results. In its simplest form, a logic model defines inputs and outputs of the program (fig. 10–3).

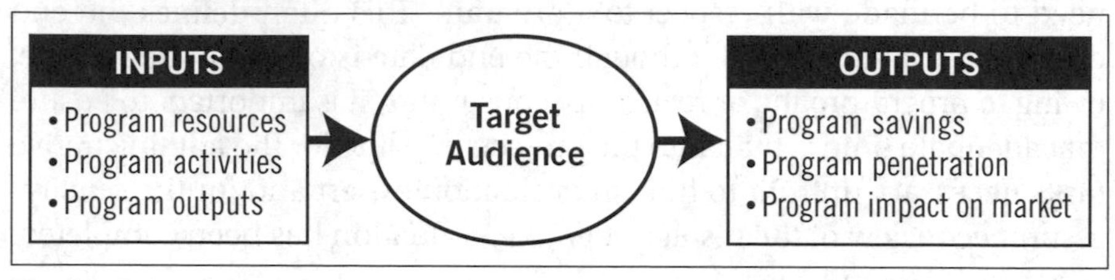

Fig. 10–3. Logic model

Logic models can be quite detailed. Well-constructed models will clarify the scope of the evaluation by explicitly identifying the problem to be addressed. The model identifies program relationships and pinpoints appropriate data to collect for assessment, aids in prioritizing questions for an evaluation to answer, and provides an understanding of the underlying theory of a program. Another purpose of logic modeling is to separate program-induced impacts from the same effects that may be generated by other factors. The ability to separate program-induced impacts from other factors becomes increasingly important as more players enter the field in which ratepayer-funded energy efficiency programs used to play alone.[7]

A key output of the design step is the development of the RFP to select an independent program evaluator. Several items must be defined and incorporated into the RFP, including the data that need to be collected, the data collection methods, the analytical methods, and the timing of the data collection. The RFP will outline the format and information required in the final report and will define the quality assurance process.

Once an evaluation contractor has been selected, there will be further refinement of the evaluation design. Highly effective evaluation contractors bring technical expertise to the table on research design methods, data collection methods, statistical sampling approaches, and analysis strategies. For example, there are numerous data collection strategies, including in-person surveys, on-site metering, and building simulations. Another consideration is whether to sample the population or to conduct a census survey that measures the entire population. The cost and the validity of the data are affected by the selected strategies; for example, on-site metering is highly accurate and reliable but also very expensive.

The final evaluation design is transformed into the statement of work for the evaluation contractor and forms the evaluation plan. This evaluation plan will be the tool used by the program administrator to manage the evaluation.

Implement the evaluation

Implementing the evaluation is the responsibility of the program administrator. The administrator will use the statement of work and the evaluation plan to manage the evaluation. Most of the work in implementing the plan lies with the evaluation contractor. The program administrator will want to establish regular progress reports with the evaluation contractor to review and manage progress.

The program administrator plays the important role of reviewing and making decisions when the evaluation does not go as planned. For example, if the contractor is not meeting established time lines, then decisions need to be made on how to expedite the evaluation or adjust the overall schedule. Quality of the data will need to be assessed, to ensure that the expectations outlined in the statement of work are met. The final output of the implementation stage is the report.

Disseminate and apply the results

The final stage is to disseminate and apply the results of the evaluation. Ideally, the evaluation was designed with the end result in mind; the results should support needed program decisions, such as whether to expand or discontinue the program. A further output is to provide the report information to key stakeholders.

To maximize the benefit of the final report, the program administrator should clearly identify and track decisions related to the report recommendations. To achieve this, program administrators may create a report that clearly outlines recommendations, decision owners, and time lines for action plans to be developed in response to the recommendations. The administrator may, in fact, create a follow-up plan that details milestones for recommendations to be acted on, implemented, and monitored for success.

The administrator will identify the various stakeholders who will have interest in the final report. Savvy administrators will create a communication plan that outlines the targeted audiences and the preferred communication channels for providing the report. For some audiences, simply providing a written report may be sufficient. Other audiences may require a meeting or meetings to review the report in detail in order to fully grasp its implications. The report may also have to be reframed for external audiences, to protect proprietary information contained in the report. This can all be identified in the communications outline.

Impact Evaluation

The impact evaluation determines the program benefits and could include any or all of the following: gross savings, net savings, nonenergy benefits, and demand savings. The gross savings are the savings at the meter, regardless of the program's influence on the actions. Net savings is the percentage of gross energy savings that is attributable to the program only. Nonenergy benefits can include a variety of attributes—typically encompassing water savings, economic impacts, and reduction in utility arrears. Demand savings measure the impact that the program has on the rate of consumption.

The challenge with an impact evaluation is that the savings cannot be measured directly. Rather, the evaluation attempts to determine the savings achieved after the program or measure has been implemented by comparing the new usage to the usage that would have occurred had there been no intervention. Hence, impact evaluations require a base case of usage.

The base case assumes that there is no intervention by a program to improve efficiency. The base case will include changes over time in underlying demographic characteristics and other factors. For example, an evaluation can analyze participants' billing data and compare their usage 12 months prior to installing the efficient equipment to their usage several months afterward to determine the level of savings achieved. The impact evaluation therefore compares the energy patterns seen today to the energy patterns that would have been expected to be seen.

Gross savings

Gross savings is "the change in energy consumption and demand that results directly from program related actions taken by energy consumers that are exposed to the program, regardless of the extent or nature of the program influence on these actions. This is the physical change in energy use after taking into account factors beyond the customers' or sponsor's control, like weather."[8] To determine gross savings, a statistically representative sample of participants are surveyed. The actual participant sites may be visited to measure and verify the savings. A statistical analysis is developed to produce program-level savings, or savings realization rates.

Realization rates are factors expressed as the ratio of the evaluation savings to the tracking savings. Tracking savings are the savings estimated at the time of the project on the basis of the efficiency measures being implemented. The evaluation savings are the savings documented by the evaluation study. For example, in a lighting project, tracking savings are based on the lightbulb replacements and the amount of hours the lightbulb is in use. The evaluation may identify that not all lightbulbs were replaced, or that the hours in use were different than noted during the implementation. Although terminology may differ among program administrators, when these realization rates are then applied to gross savings, the savings are generally known as *adjusted gross savings.*

For example, at NSTAR, we evaluated the Energy Star lighting program. One hundred participants were randomly selected to participate in the study. Site visits were conducted to verify the number of lightbulbs in use. Additionally, a light logger was used to determine the number of hours per day per room the lights were in use. Interestingly, the measurements obtained using the light loggers revealed that, when asked how many hours the lights were in use, customers tended to overestimate the hours of use (fig. 10–4).

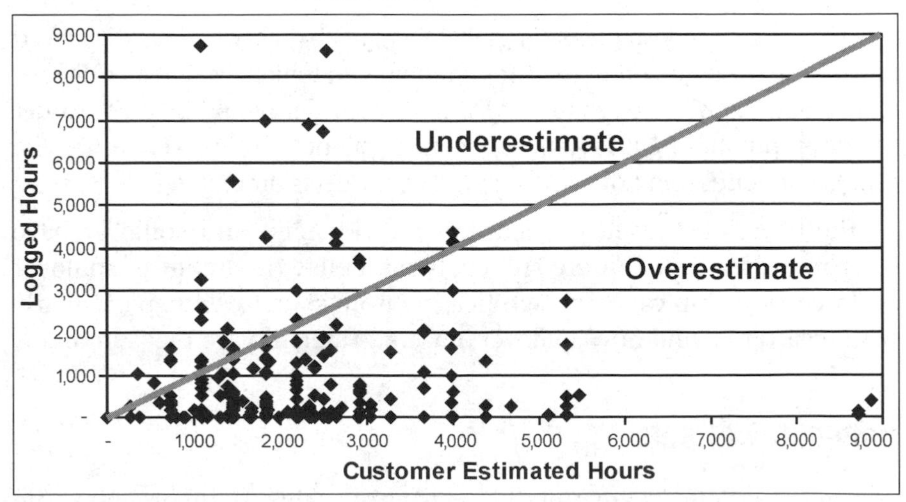

Fig. 10–4. Example output from impact evaluation on lighting—customer estimated versus logged hours

Net savings

The net savings or impact is the percentage of gross energy impact attributable to the program.[9] The formula for calculating net savings starts with gross savings adjusted by the realization rate (i.e., adjusted gross savings) and adds spillover savings while negating free-ridership savings:

$$\text{Net savings} = \text{gross savings} \times \text{realization rate} \times (1 - \text{free-ridership} + \text{spillover})$$

The realization rate is an impact factor, expressed as the ratio of the evaluated savings to the tracking savings. The closer this value is to 1.0, the more accurate the tracking system is at capturing savings.

Spillover comprises additional measures implemented by the program participant that were induced because of participation in the program. For example, in the Energy Star lighting program NSTAR evaluated, 12 CFLs were provided and installed as part of the home energy audit program; if, in the sample of 100 participants, some participants were found to have installed additional CFLs on their own after seeing the program benefit, thus increasing the number in the household, this additional savings would be considered as spillover.

Additionally, spillover, as described in the preceding example, may be participant spillover or nonparticipant spillover. Nonparticipant spillover comprises actions influenced by an energy program but not linked to participation. With lighting, increased awareness influenced by marketing, may result in marketplace changes like stocking practices.

Free-ridership constitutes savings that the participants would have achieved in the absence of the program through their own initiatives or expenditures. Consider an appliance rebate program. A free rider would be a consumer intent on purchasing an energy efficient appliance. To this consumer, the rebate is a bonus and not a factor in his or her decision making.

The techniques used to measure free-ridership and spillover have improved over time but are still evolving. Policy treatment of spillover and free-ridership can vary significantly from state to state, particularly with regard to if and how spillover and free-ridership are measured.

Nonenergy effects

Nonenergy effects are other impacts attributable to the program. The definition of nonenergy effects depends on public policy and varies from state to state. Nonenergy effects can be positive or negative and are segmented into resource and nonresource effects.

Resource effects are linked to reduced usage of other resources, such as water, gas, or oil. Nonresource effects, such as reduced arrears, are items not associated with a resource and are often more difficult to quantify. Items frequently included among nonenergy effects are air-emissions benefits, water savings, labor savings, increased revenue, increased comfort, and reduced maintenance costs. Water savings can be achieved, for example, in commercial kitchen programs featuring low-flow sprayers. Reduced arrears and write-offs are benefits often associated with low-income programs, in which lowered energy use results in lower bills and, hence, lower arrears and potential write-off.

While nonenergy effects are usually benefits, they can also be negatives. An example of a negative resource is the heating penalty that a customer will experience when changing from incandescent lamps to fluorescent lamps when they are installed in a heated space. The heat that was being generated by the incandescent lamp will now need to be generated by their fossil fuel heating source (oil or natural gas), resulting in an increase, not a decrease, in the nonelectric energy use.

Demand savings

There is increased interest in evaluating the impact that EE programs have on the rate and timing of consumption, or the demand. This is particularly applicable to demand response programs but is also important in energy efficiency programs. The advent of markets, such as New England's Forward Capacity Market (FCM), in which energy efficiency portfolio administrators can bid their demand savings into the market is driving interest in accurately calculating demand savings.

Protocols associated with determining demand savings for energy efficiency have some additional requirements. These protocols require data collection and analysis during specific periods. For example, if the peak demand is during the summer, from 1:00 p.m. to 6:00 p.m., then the rate of consumption during that time needs to be determined. With the introduction of more sophisticated metering capable of collecting usage by time, the data for calculating demand savings have become more accessible.

Several demand calculations are completed during evaluations, including annual average demand savings, peak demand reductions, coincident peak demand reductions, and demand response peak demand reductions. The average annual demand savings is the annual energy savings divided by the number of hours in a year (8,760 hours). The peak demand reductions involve determining the maximum amount of demand reduction during a period of time (i.e., a season). Coincident peak demand reductions are demand savings that occur when the load service utility is at its peak demand for all customer segments. Calculating this requires data on the utility's peak demand periods.

Other Evaluation Types

In addition to impact evaluations, the other evaluations used are process, market-effects, and cost-effectiveness evaluations. Process evaluations are designed to identify improvements in the delivery or performance of a program. Market-effects evaluations seek to understand the program's impact on the overall supply chain. Cost-effectiveness evaluations quantify the cost of the program and compare those costs to the benefits of the program.

Process evaluations

Process evaluations are valuable to identify performance gaps of the program. Program administrators are keenly interested in where costs can be removed from the program and where quality of service can be improved. Process evaluations involve gathering information from all stakeholders in the delivery of a program, including program designers, implementers, participants, policy makers, and trade allies. Areas for improvements in efficiency and effectiveness and in program delivery are identified. For example, a process evaluation may identify how to reduce delivery time to participants or whether the program incentive is appropriate, too high, or too low.

Market-effects evaluations

Market-effects evaluations attempt to understand the impact that the program has on the market and how the market is changing. As end-use technologies progress through the market adoption cycle, a market-effects evaluation can determine when, for example, transformation may occur. As Edward Vine, of Lawrence Berkeley National Laboratory, has observed, "A key market indicator in the markets effects evaluation of EE [energy efficiency] programs is the value of energy savings from sales and/or market share changes for targeted efficiency measures. Other related indicators include changes in awareness, intention to purchase, stocking practices, product availability, prices and willingness to invest in EE."[10]

A form of a market-effects evaluation is a *market potential* evaluation. In this type of evaluation, the evaluator attempts to quantify the potential savings over time for a program. This involves determining program baselines and understanding potential penetrations of various end-use technologies and customer segments. This is an important tool for program administrators developing multiyear programs.

Cost-effectiveness evaluations

Cost-effectiveness evaluations are key components of program development. There are several cost-effectiveness tests that calculate the program cost and the program savings. Which test or tests may be used is driven by policy, which varies from state to state. In fact, over one-third of states require multiple tests.[11]

For program administrators, the cost-effectiveness tests are key components in assessing individual programs and building a program portfolio that will be approved by regulators. Program administrators will want to demonstrate fiscal responsibility in delivering programs that maximize the benefits and minimize the costs.

The benefits associated with a program include the energy, demand, nonenergy benefits, and externalities. Externalities are similar to nonenergy benefits and include items such as reduced carbon and sulfur emissions. Which types of cost-effectiveness tests are used often depends on public policy. The costs associated with a program include administration, implementation, marketing, and evaluation costs, although, depending on the test, they may include impact on rates and customer costs. Table 10–2 provides an overview of the cost-effectiveness tests, the measurement approach, general costs, and benefits included.

Utility cost test. This test looks at the costs of an EE program through the lens of a utility and therefore includes only the utility costs; this test does not include the participant costs. The benefits to the transmission and distribution system, along with energy and demand, are included; nonenergy benefits are not included in this test.[12]

Program administrator cost test. This test looks at the cost through the lens of a program administrator. The costs included are the program administrative costs, the incentives, and the costs associated with supply. The benefits to the transmission and distribution system, along with energy and demand, are included.[13]

Table 10–2. Attributes of cost-effectiveness tests (Source: Titus et al., 2004)

Test	Measurement Approach	General Cost Included	General Benefits Included
Utility test[1,2]	Measures net costs taking perspective of utility; excludes participant costs	Utility costs	Avoided supply, transmission and distribution, generation and capacity costs during load reduction periods
Program administrator cost test[2]	Measures net costs based on administrative costs only	Program administrative costs; incentives; increased supply costs during periods of increased load	Net avoided supply costs; marginal cost of reduction in transmission and distribution, generation, and capacity during load reduction periods
Participant test[1,2]	Measures quantifiable costs and benefits taking customer perspective	Expenses incurred by customers, increases in customer utility bills, value of customer time spent arranging program participation	Reduction in customer utility bills, incentives paid, tax credits, gross energy savings
Ratepayer impact measure, or nonparticipant test[1,2]	Measures program impacts on customer bills or rates	Initial and annual program costs incurred by administrator and any other parties, incentives paid, decreased revenue from load reduction periods, increased supply costs from load increase periods	Savings from avoided supply cost, including transmission and distribution, and generation; capacity cost reductions during load reduction periods; increased revenue during load increase periods
Total resource cost test[1,2]	Measures net costs taking perspective of utility, but includes participant and nonparticipant costs; applied at program and/or measure level; usually focuses on measures or activities for a single year	Program costs paid by utility and participants; increase in supply costs during load increase periods; spillover	Avoided supply costs; reduction in transmission and distribution, generation, and capacity costs; tax credits
Societal test[1,2,3]	Based on TRCT, but takes perspective of society. Applied at program and/or measure level. May use higher marginal costs than TRCT; should use societal discount rate; excludes tax credits & interest	All costs included in TRCT, plus: externalities, some non-energy costs (including costs to participants and society)	All benefits included in TRCT, plus externalities (avoided environmental damage, increased system reliability, fuel diversity) and some nonenergy benefits (including benefits to participants and society)
Public purpose test[1,3]	Based on societal test, taking societal perspective, and taking long-term view; applied at portfolio level	Same as societal test, but takes into account market effects and broader array of externalities and nonenergy costs	Same as societal test, but takes into account market effects and broader array of externalities, nonenergy benefits; spillover savings

[1]Sebold (2001). [2]California State Governor's Office (2001). [3]TecMarket Works Framework Team (2004).

Participant test. The participant test assesses the program through the lens of the participants. In this case, the only costs included are the customer costs; this includes the expenses incurred by the customer to participate in the program, and a value is placed on the customer's time in arranging for the work to be completed. The benefits, from a consumer or participant perspective, are the energy and demand savings to the consumer, along with other nonenergy benefits.[14]

Ratepayer impact measure. This test attempts to understand the impact on ratepayers' or consumers' bills as a result of implementing programs. This test looks at the impact of program costs applied to all utility consumers, owing to changes in utility revenues and operating costs. All program costs, both initial and ongoing, are included. This measure also accounts for lost revenue associated with reducing demand and gives credit for increased revenue if appropriate. The supply works similarly where costs and benefits are monetized and included in this measure. Savings from a transmission and distribution perspective are also determined. The major component of savings is associated with energy and demand.[15]

Total resource cost test (TRCT). The TRCT measures the net cost of an EE program as a resource option, based on the total costs, including utility and participant costs. The benefits include energy, demand, transmission and distribution, and both resource-based and nonresource-based nonenergy benefits. This is the most commonly used test.[16]

Societal test. The societal test is a version of the TRCT that goes beyond the lens of the utility to look at the benefit to society as a whole. The difference between this test and the TRCT is that this test includes benefits for externalities, such as carbon and sulfur emissions, health care costs, and tax benefits.[17]

Public purpose test. This test is applied at the portfolio perspective and takes a long-term view. The costs and benefits are similar to the societal test, market effects and a broader array of externalities and nonenergy costs and benefits are taken into account.[18]

Summary

Evaluation is a vital element in the optimization of a EE program. From a program perspective, evaluation feeds the determination of program objectives and design; it also feeds into policy objectives and aids in ensuring that these objectives are being achieved. The role that evaluation plays in resource planning, while often overlooked, is increasingly being regarded as important. From a resource-planning perspective, evaluation helps to determine the extent to which energy efficiency resources can be substituted for supply sources.

Impact, market effects, cost-effectiveness, and process evaluations provide insight into the performance of the programs and identify opportunities for program administrators to improve efficiency and effectiveness. Additionally, evaluations are a fundamental tool used by program administrators to justify continued investment in EE and to ensure cost recovery for programs delivered.

Savvy program administrators are fully leveraging evaluation to optimize the performance of their programs, to aid in resource planning, and to demonstrate the benefits of the programs to the various stakeholders. In the words of Monica Nevius, Consortium for Energy Efficiency's senior program manager for research and evaluation, "Evaluation is a science that continues to improve in its methods and accuracy. At the same time there is increased interest and demand for EE programs to perform at unprecedented levels. Program administrators who are successful in the long term at delivering the most value for their programs will maintain an expertise in evaluation techniques and continue to push the science for improvement."[19]

References

1 National Action Plan for Energy Efficiency. 2006.

2 Barnes, Harley, and Gretchen Jordan. 2006. *EERE Guide for Managing General Program Evaluation Studies.* Washington, DC: U.S. Department of Energy, p. 18.

3 Nevius, Monica. Personal communication with Penni McLean-Conner on August 4, 2008

4 Vine, Edward. 2007. The integration of energy efficiency, renewable energy, demand response and climate change: Challenges and opportunities for evaluators and planners. Presented at the Energy Program Evaluation Conference, Chicago.

5 Ibid.

6 Barnes and Jordan, 2006.

7 Consortium for Energy Efficiency. Energy Efficiency Program Evaluation: A Guide to the Guides. http://www.cee1.org/eval/eval-res.php3

8 Schiller, Steven R. 2007. *Model Energy Efficiency Program Impact Evaluation Guide,* p. 3-4. http://www.epa.gov/cleanenergy/energy-programs/napee/index.html

9 Ibid.

10 Vine, 2007.

11 Titus, Elizabeth, Monica Nevius, and Julie Michals. 2004. *How Do We Measure Market Effects? Counting the Ways, and Why It Matters. In Proceedings of the ACEEE 2004 Summer Study on Energy Efficiency in Buildings.* Washington, D.C.: American Council for an Energy-Efficient Economy ACEEE.

12 Ibid.; Schiller, 2007.

13 Titus et al., 2004.

14 Ibid.

15 Ibid.; Schiller, 2007.

16 Schiller, 2007.

17 Titus et al., 2004.

18 Ibid.

19 Nevius, Monica. Personal communication with Penni McLean-Conner on August 4, 2008

Chapter 11

Positioning for the Future

"We are now passing the tipping point on climate change, as uncertainty has turned into consensus, and talk is turning into actions."[1] Clearly, EE is one of the solutions to address climate change. Today there is a demand for new solutions to meet growing energy demands. In contrast to traditional, predominantly fossil-based generation, the 21st-century solutions of EE, DSM, and renewable energy support a sustainable environment.

Many states are leading the charge with the passage of groundbreaking legislation and regulations. In 2008, Massachusetts, for example, passed the Green Communities Act of 2007, which directs utility program administrators to secure all energy efficient alternatives that are cheaper than supply; on this foundation, the Massachusetts public service commission issued an order on decoupling. Combined, these actions create an environment to meet growth in energy with demand-side and renewable options in a way that also provides a positive cost-recovery environment for utilities.

Legislation and regulation across the country is resulting in the rapid launch or expansion of EE programs. As such, EE leaders are working to create and build an EE culture grounded in a solid business case and supported with appropriate policy and regulation. EE leaders are also delivering quality EE programs and are actively looking to optimize the programs to meet increased savings goals. Finally, EE leaders recognize that it is not enough to focus year to year, but rather, to be successful in the years to come, it is critical to plan several years in advance to position the programs from a people, process, and technology perspective.

From a people perspective, EE program administrators are focused on establishing a sufficiently trained workforce to design and deliver quality programs. Former New York State Energy Research and Development Authority president and chief executive officer Paul D. Tonko noted that "Educating and training a workforce is vital to [New York's] efforts to continually make investments in clean energy technologies."[2] Investments in clean energy technologies are expected to produce rapid and significant growth. In energy efficiency alone, the forecast is for rapid expansion, at 15% per year.[3] EE program administrators positioning for the future realize that rapid expansion can be constrained by the lack of a trained workforce to deliver programs. Therefore, program administrators are already putting in place infrastructure that supports the much anticipated training and development of human resources to support expanding EE efforts.

From a process perspective, program administrators are interested in enhancing their program design and delivery processes to reduce costs and improve effectiveness. Administrators today are proactively discussing how to establish integrated delivery of EE and DSM programs, where the program provides not only energy efficiency solutions but also demand response and renewable distributed-generation solutions. The demand for maximized energy savings requires innovation of program delivery and participation. Administrators are redesigning programs to reach untapped audiences such as non–English-speaking customers through multicultural outreach activities. While energy savings have been and will continue to be based on the application of more efficient end-use technology, there is also an increased focus on behavior-based programs, which encourage consumers to change their behaviors and thereby reduce energy usage.

From a technology perspective, a major shift is occurring as utilities begin to replace—or at least consider replacing—traditional meters that provide monthly readings with advanced metering infrastructure (AMI) that enables the utility and the consumer to receive real-time energy usage information and pricing signals. EE leaders are evaluating how to leverage and integrate technology with efficiency and behavior-based real-time pricing programs. This evolution of technology has implications to EE programs.

People

There is excitement among energy service professionals about the growth in careers related to energy efficiency and renewable energy—sometimes referred to as *green-collar jobs*. Green-collar jobs are "paid positions providing environmentally-friendly products or services."[4] These new skilled jobs range from installing solar panels, to retrofitting old buildings with energy efficient and renewable solutions, to designing and delivering programs to aid consumers in managing energy. Excitement about growth in green-collar jobs is, however, tempered by the recognition that this growth may lead to a shortage of available and experienced personnel for energy services.

Energy service leaders interested in successfully positioning their programs for the future are actively involved in workforce planning. Workforce planning is a key component to ensure that a sufficient and qualified workforce is ready to fill the new green-collar jobs associated with EE and other clean energy solutions.

The Corporate Leadership Council has defined *workforce planning* as a "systematic assessment of workforce content and composition issues and determines what actions must be taken to respond to future needs." Workforce planning is grounded in a needs assessment. A workforce plan will identify solutions to bridge the gap between the needs of the current and the future workforce.

The needs assessment is the starting point for workforce planning. A needs assessment seeks to understand the future state of workforce requirements along with the capabilities and skills needed. The current state of the workforce is also documented and a gap analysis is completed to compare the current and future states of the workforce. The needs assessment will also provide recommendations on educational and training capabilities that will be needed to address this gap.

Initiatives to address the gaps from the needs assessment will vary depending on the location, expected growth of green-collar jobs, and specific skills and capabilities needed. A variety of approaches are being used, including partnering with training institutions, establishing scholarships, and developing co-op or intern programs.

Partnering with local training institutions is a proven workforce sourcing strategy. The partnership involves identifying skills, developing curriculum, and providing training. Program administrators may partner with local training institutions such as technical high schools, community colleges, and universities.

Offering scholarships is another tool that increases awareness of the opportunities associated with green-collar jobs, and also serves to build the workforce pool. The scholarships can support a specific course of study or fund attendance at industry events or conferences. Industry organizations or program administrators may find value in offering scholarships as a means to jump-start interest in green-collar jobs. Providing students with scholarships to attend an industry event exposes them to other professionals and provides them with insight into the issues the industry is addressing in real time.

Interns or co-op students represent another excellent sourcing strategy. Many companies are leveraging this recruitment method, as evidenced by the placement annually of over 250,000 U.S. students.[5] Cooperative education enables students to apply their technical knowledge, gain an understanding of professional and ethical responsibility, and practice effective communication. Research on cooperative education indicates that co-op students make more informed career decisions, have higher grade-point averages, and are more successful in early socialization and adjustment to a company.[6]

Process

Improving the processes associated with EE programs is an ongoing effort. EE leaders are working together in various organizations to share ideas. Participating in industry organizations is one of the best ways for EE administrators to keep up with the latest developments and ideas. EE administrations will benefit from participating in these industry organizations with techniques to continually improve the process of program delivery. From a high-level perspective, there are three constructs that offer the opportunity to dramatically change EE delivery: integrated program delivery, multicultural outreach, and behavior-based programs.

Integrated program delivery

Consumers interested in reducing their energy usage do not think of energy usage in categories of energy efficiency, demand response, and renewable energy. Nor do they think in terms of electric and gas programs. Rather, they look to utilities and program administrators to offer services that maximize their opportunities for EE solutions. Today, though, many programs have a singular focus. For example, there may be a program on gas energy efficiency from one source and a program on electric energy efficiency from a different source. Further, consumers interested in both solar panels and reducing energy waste will often have to identify and coordinate those solutions on their own.

A solution to this singularly focused delivery is to integrate the delivery of the programs from a fuel and program-type perspective. Program administrators considering integrated program delivery must redesign their program delivery and partner with others as appropriate.

Customer outreach needs to be designed to focus on integrated total energy solutions. This implies a broader message focused on how consumers can make lifestyle changes to support a greener environment by taking action with respect to holistic demand-side options. Market assessments will help administrators to better target customers interested in total integrated solutions.

Program administrators need to consider how to adjust the customer intake or screening methods. Typically, administrators screen applicable customers over the telephone or through an on-site audit. With integrated delivery, program managers must identify additional screening tools that assess a customer's applicability for demand response and on-site renewable solutions, in addition to energy efficiency programs. Designing the support assessment tools to complete this screening and training the customer service representatives and on-site auditors are some of the changes that must be implemented to support integrated program delivery. There is an incremental cost associated with this screening that must be understood and documented for program design and evaluation.

Delivery of an integrated program requires a program manager to line up multiple program vendors. Specialists in the installation of solar panels may not be qualified to enhance building insulation or retrofit lighting systems. A program manager in the design of the process must identify and secure program implementers for the various solutions and provide training in the design stage of the program implementation process. Most important, the program manager needs to ensure that from the customer perspective, the implementation is seamless.

An example of integrated delivery is the Marshfield, Massachusetts, pilot program being implemented by NSTAR. This is one of the first programs in the nation to offer integrated program delivery. The goal of the program is to address growing energy demand with demand-side resources, as opposed to the traditional method of building additional capacity. In the Marshfield program, an engineering review by the utility indicated that capacity upgrades are needed to support additional load growth of 2 MW on a 25 MW current load. The goal of the Marshfield program is to secure 2 MW of demand-side solutions to delay the needed capacity upgrades. The program consists of applying the demand-side solutions of energy efficiency, demand response, and on-site renewable solar generation.

The program manager for the Marshfield pilot faced limited information on design parameters for an integrated program. Hence, the program manager started with a charrette to develop the design. (A charrette is a collaborative session in which a group of designers draft a solution to a design problem.) In the Marshfield pilot program, the charrette included local business leaders, energy service companies that serve as program implementers, EE experts, and regulatory experts. In the case of the Marshfield project, the charrette was formed to gather input that was used in the design of an integrated energy efficiency, demand response, and solar (photovoltaic) program. The outcome of the charrette was not only the assembly of great ideas on integrated program design but also the creation of joint ownership of the solution among the stakeholders participating in the charrette.

Ultimately, the charrette also helped with the design of the marketing approach, which focused on consumer lifestyles. The Marshfield project uses a community-based marketing approach, the slogan of which is reproduced in figure 11-1.[7] The marketing tools used to spread the message of the program to the community included outreach to local community leaders, participation in local events, and targeted marketing in Marshfield's local newspaper.

Fig. 11–1. Marketing material for Marshfield (MA) integrated EE, DSM, and renewable energy program

The customer assessment occurs through the home or commercial energy audit. The auditors are trained to assess not only energy efficiency but also demand response and photovoltaic opportunities. At the conclusion of the audit, the consumer is presented with a package of solutions that include energy efficiency, demand response, and photovoltaic.

Delivery of the actual measures is completed through program implementers. Demand response measures can be installed during the home energy audit, and a follow-up for additional efficiency measures can be scheduled during the audit. If the customer is a good candidate for solar measures, a follow-up appointment with the solar installation partner is scheduled. At this session, further assessment of applicability occurs and a contract is presented. Early results from the program are strongly positive. The two-year program has already completed 25% of planned energy audits in the first three months of operation. Lead generation for photovoltaic is resulting in a 45% closure rate, significantly higher than traditional results using a stand-alone approach.

Multicultural outreach

As program administrators face increased goals for energy savings, they are looking at customer segments that have been largely untapped. Multicultural outreach is one approach to targeting customers for whom English is not their first language.

Multicultural marketing is defined as targeting and communicating to ethnic segments by recognizing their cultural framework. The population of customers for whom English is not their first language is growing. Over 11% of the U.S. population is born outside the United States, and nearly 1 in every 10 of the nation's counties has a population in which multicultural groups comprise more than 50% of the total.[8] The reason program managers are investing in multicultural marketing is that it helps reach underserved customer populations and increases customer satisfaction with the company.

Designing and delivering programs to reach these customers means reevaluating marketing outreach and program delivery. Best practices for multicultural outreach include communicating with customers by using multiple channels in the appropriate language and partnering with community-based organizations to reach this audience.

Providing customers for whom English is not their first language with communication channels that are designed for them serves as a foundation for reaching this audience. The channels include the call center, outside events, direct mail, and the Web site. Where appropriate, program managers will want to ensure that the translation of the materials is accurate and reflects cultural differences.

Partnering with community-based organizations is a powerful tool for implementing EE programs. A community-based organization can identify customers, provide customer intake and translation services, and support publicizing the program. Program managers that have had success with multicultural outreach and marketing have found that it is important to create a positive partnership with community-based organizations and leverage their input and advice on how to deliver the program.

Behavior-based programs

While programs centered on end-use technology replacement are the most common, more program managers are augmenting their program portfolio with programs targeting customer behavior, such as Progress Energy Florida's Save the Watts program. Jeff Lyash, president and chief executive officer of Progress Energy Florida, noted that "We have some of the nation's most successful and innovative energy efficiency programs, and changing consumer behavior is key to their success."[9] Programs affecting customer behavior strive, through education and customer communications, to encourage customers to change lifestyles or behaviors to reduce energy. Behavior-based actions range from turning the thermostats down in winter, to unplugging appliances when not in use, to simply turning off unneeded lights. Behavior-based programs often include Web-based tools, real-time monitoring tools, and real-time price responsiveness programs.

The World Wide Web is a growing and very cost-effective channel to provide customers with energy management tools that target customer behavior. Program administrators are offering educational tools that are commercially available, along with custom designed tools they have developed in-house. Real-time monitoring and analysis tools offered on the Web are becoming more popular. Energy analysis software provides customers with tools that allow them to understand where and when they are using energy so that they may better manage their usage and reduce costs. These tools allow the customer to receive energy data daily, or in real time, in a user-friendly format. With more timely usage information, customers are better able to understand their load profiles and reduce usage during periods of high prices.

Another real-time monitoring tool is a PowerCost Monitor. This type of tool displays a home's electricity consumption in real time, in dollars and cents, and in kWh through a user-friendly tabletop display or a Web connection. The premise behind the technology is that when customers can see what they are spending on electricity in real time, they have a higher propensity to cut back on electricity consumption, thereby saving energy and money on their monthly bills.

Real-time pricing and critical peak pricing represent another form of behavior-based programs. These offerings are most successful when partnered with energy efficiency solutions. These offerings allow prices to be adjusted frequently to reflect real-time system conditions.

Combining price signals with control technology can significantly influence customer behavior and increase response to demand events. The California Statewide Pricing Pilot showed a 50% increase on peak period impact for residential customers with control technology and pricing signals. This held true for commercial and industrial customers: a 13% reduction occurred on critical peak days with control technology and pricing signals, while there was no reduction for similar customers without these drivers.[10]

Gulf Power, who has offered a price response load-control program for several years, indicates that customers see approximately a 10% savings. The installation of AMI, enabling real-time pricing signals and demand response, will continue to advance the design and delivery of behavior-based programs.

Technology

As program administrators position for the future, they must plan on integrating programs with AMI investments. Chartwell data indicate that, in fact, "over half of the utilities are either installing or seriously assessing advanced metering on some level."[11] AMI goes beyond the meter and includes communications systems used to notify the customer of a critical peak and, in some cases, enables communication within the customer's premises to manage equipment.

Utilities can and have been delivering incentive-based demand response solutions and time-based pricing by use of existing customer information systems. The interest in AMI stems from two sources: first, a concern that the piecemeal approach of delivering demand response through existing systems and stand-alone technology is limiting; and second, the opportunity a more robust AMI platform brings to other applications, in addition to the other associated utility benefits.

As a result, utilities are engaged in serious discussion on the role and implications of AMI. AMI includes systems that measure, collect, and analyze energy usage from advanced devices—such as electric meters, gas meters, and/or water meters—through various communications media on request or on a predefined schedule. This infrastructure includes hardware, software, communication devices, customer-associated systems, and meter data management software.

AMI consists of a robust two-way communications infrastructure and uses the meter as the gateway device to the premises. AMI has three characteristics:

- Solid-state or computerized meters that collect time-series (interval) energy-use data and are programmable to support features such as time-of-use rates
- Two-way communication between the meters and the utility
- The ability to support applications beyond meter reading, such as demand response, outage communication, and remote connect and disconnect

To support AMI, a meter data management system is needed to collect, analyze, manage, and store the volumes of data associated with the capture of interval energy usage. Communications systems are also needed. Most utilities believe that multiple communications gateways may be needed to communicate with every meter. These gateways include radio frequency, power line carrier, wireless, and broadband over power line. While a utility may have a preferred method for communicating, there may be some unique metering locations, requiring customized solutions using other gateways.

Communication to the customer about critical peak periods is needed to drive customer behavior. This communication may be achieved through Web portals, e-mail and text messaging, bulk outgoing telephone messages, and even control signals sent directly to devices in the home.

Finally, communications systems are needed within the consumer's home to control equipment usage during periods of high demand to fully support customer behavior. This communication can be achieved through home area networks. Home energy networks are in-premise, low-data networks between the meter and equipment inside the home, such as thermostats and pool pumps. Achievement of home area networks requires that a gateway, which could be the meter, and appliances be equipped with a communications interface. Today most appliances do not come equipped to "hear" a signal from the gateway.

Harvey Michaels, founder of Nexus Energy, has predicted that someday consumers will be able to set their home for ultraefficiency on critical peak days; and this setting will automatically cascade into signals and operation of appliances that minimize usage. Other consumer-friendly settings predicted by Michaels are holiday or visiting-relative settings. With growing interest and investment by utilities and industry in the smart grid, Michaels' predictions may be realized in the not-too-distant future.

AMI will provide consumers with better control over their energy consumption. For utilities, it will provide more accurate and timely information, which in turn will support more accurate billing and financial statements. Customer service will be supported with this infrastructure as faster—and, in some cases, immediate—service can be provided, as in the case of a connect order, which, with AMI, can be done without field involvement. For DSM program administrators, AMI offers a new platform on which to design and deliver programs.

Summary

Creating and building upon an EE culture is a continual process. Even utilities that have been engaged in EE for a long time are facing dramatic changes to fundamental business models as the concept of decoupling is advanced. Regulatory bodies are propagating new rules and rate models that position utilities to actively and enthusiastically engage in EE. To gain internal and external support and alignment, program administrators must start with a solid business case for EE investment.

Program administrators must not only work to deliver EE programs but also strive to enhance and optimize the programs. Evaluation plays a key role in identifying process improvements. Participating in industry organizations serves to advance policy, offer professional development, advance programs, and provide research and development and education.

Ultimately, optimizing EE programs means successfully positioning these programs for the future. Successful EE leaders will look toward the latest people, process, and technology trends and ensure that programs and staffing are positioned to leverage the trends.

From a workforce perspective, the good news is that this is a growing field. However, that also presents a challenge for program administrators, who must rapidly expand their staffing to support expanded programs. Program administrators with an eye toward the future are working jointly with universities and technical schools to develop curriculum for energy management jobs.

From a process perspective, reaching markets that have been largely untapped requires new program design and delivery. Integrating delivery of EE, DSM and renewable energy programs provides benefits to consumers, such as one-stop shopping, and also provides benefits for program administrators, by reducing costs associated with the program delivery. Behavior-based programs are also expanding and require new models.

Finally, to successfully position programs for the future, program administrators must consider the evolution of the metering infrastructure. Advanced metering opens up new opportunities for behavior-based programs combined with real-time information and price signals. With the technology evolution occurring from a metering infrastructure perspective, new opportunities will leverage this infrastructure.

Applying the principles and fundamentals of EE will provide long-term benefits for customers, utilities, and the environment. These principles and fundamentals rely on creating an advanced EE culture, delivering EE, and optimizing the programs.

References

1 Accenture Executive Survey on Climate Change. 2008. Rising to the challenge—and seizing the opportunities.

2 New York State Energy Research and Development Authority. 2008. Clean energy workforce training initiative and proposed wind energy research and testing center to be established in New York State. News release, February 25. http://www.nyserda.org/press releases/2008/pressrelease20082502.asp

3 Goldman, Chuck. 2008. Energy efficiency services workforce needs assessment. Presented to NAPEE Leadership Group Meeting.Scottsdale, AZ

4 Ella Baker Center. Green collar jobs campaign glossary of "green wave" terms. http://www.ellabakercenter.org/page.php?pageid=29&contentid=32

5 Pettit, D. E. 1998. *1998 Census of Cooperative Education Executive Summary.* MD: The Clearinghouse for Cooperative Education.

6 Brows, S. 1985. The relationship of cooperative education to organizational socialization and sense of power in first job after college. Doctoral dissertation, Boston College; Edison, K. 1981. Cooperative education and career development at Central State and Wilberforce Universities. Doctoral dissertation, Harvard University; Gardner, Philip, and Steve W. J. Koslowski. 1998. Learning the ropes: Coop students do it faster. *Journal of Cooperative Education.* 28: 30–41; Gardner, Philip D., David C. Nison, and Garth Motschenbacker. 1992. Starting salary outcomes of cooperative education graduates. *Journal of Cooperative Education.* 27: 16–26; Gillin, L., R. Davie, and K. Beissel. 1984. Evaluating the career progress of Australian engineering graduates. *Journal of Cooperative Education.* 23: 53–70; Mann, R., and D. Schlueter. 1985. The relationship between realistic expectations about work and the cooperative education work experience. Boston: Northeastern University, Cooperative Education Research Center; Patterson, Valerie. 1997. The employer's guide: Successful inter/co-op programs. *NACE's Journal of Career Planning and Employment.* 57 (http://www.naceweb.org/pubs/journal/wi97/patterson.htm); Pettit, 1998; Pittenger, K. 1993. The role of cooperative education in the career growth of engineering students. *Journal of Cooperative Education.* 28: 21–29; Van Gyn, G., J. Cutt, M. Loken, and F. Ricks. 1997. Investigating the education benefits of cooperative education: A longitudinal study. *Journal of Cooperative Education.* 32: 70–85; Wessels, W., and G. Pumphrey. 1995. The effects of cooperative education on job search, quality of job placement, and advancement. *Journal of Cooperative Education.* 31: 42–52.

7 Marshfield Energy Challenge, marketing collateral; NSTAR, May 2008.

8 American Community Survey; AMACOM.

9 Lyash, Jeff. 2007. Progress Energy Florida kicks off customer education program. New release, June 28. http://progress-energy.com/aboutus/news/article.asp?id=16422

10 George, Stephen. 2007. Enabling the 21st century utility. Presented at the Nexus Energy Software Client Conference; Scottsdale, AZ.

11 Chartwell Inc. 2008. Smart Grid: *How Utilities View the Grid of the Future.* Chartwell Metering Research Series. Chartwell Inc., p. 16.

Index

A

B

C

D

E

F

G

H

I–J

K

L

M

N

O

P

Q

R

S

T

U

V

W

X–Z